Reactivity and Structure
Concepts in Organic Chemistry

Volume 27

Editors:

Klaus Hafner   Jean-Marie Lehn
Charles W. Rees   P. von Rague Schleyer
Barry M. Trost   Rudolf Zahradník

Steven V. Ley · Caroline M. R. Low

Ultrasound in Synthesis

With 23 Figures

Springer-Verlag
Berlin Heidelberg New York
London Paris Tokyo Hong Kong 1989

Steven V. Ley · Caroline M. R. Low

Imperial College of Science and Technology
Department of Chemistry
South Kensington
London SW7 2AY, UK

ISBN-13:978-3-642-74674-1   e-ISBN-13:978-3-642-74672-7
DOI: 10.1007/978-3-642-74672-7

Library of Congress Cataloging-in-Publication Data
Ley S. (Steven), 1945
Ultrasound in synthesis / S. Ley, C. Low. p. cm. –
(Reactivity and structure: concepts in organic chemistry; v. 27)
1. Chemistry, Organic-Synthesis. 2. Ultrasonic waves. I. Low, C. (Caroline), 1962.
II. Title. III. Series: Reactivity and structure; v. 27.
QD262.L45   1989   547'.2–dc20   89-10099
ISBN-13:978-3-642-74674-1

© Springer-Verlag Berlin Heidelberg 1989
Softcover reprint of the hardcover 1st edition   1989

2151/3020-543210 – Printed on acid-free paper

List of Editors

Preface

The effects of heat and light on chemical reactions have long been known and understood. Ultrasound has been known to promote chemical reactions for the past 60 years, but despite this, it did not attract the attention of synthetic chemists until recently. This arose historically from early studies which concentrated almost exclusively on reactions in aqueous media and was also, in some measure, due to the availability of suitable technology. Since the early 1980s a plethora of literature has appeared of direct interest to synthetic chemists and the field has been developing rapidly.

The aim of this book is to bring the background of this fascinating field to the attention of a wider audience. It explores the literature to date and attempts to indicate other areas in which ultrasound may be exploited. It also hopes to explode some of the myths surrounding this area which have hitherto been regarded by the synthetic community as a bit of a black art!

Existing books and reviews have tended to concentrate on the physics of sonochemistry and to catalogue the instances in which ultrasound has proved useful in tackling synthetic problems. Our aim has been to stress the relevance of this technique to synthetic chemists and we have included a section which deals with the practical aspects of carrying out these reactions. We have also given an indication of the instrumentation that is currently available, although it should be stressed that the majority of the reactions described were carried out using nothing more complex than a standard cleaning bath of the type that can be found in most laboratories.

The range of reactions described is extremely diverse and extends from the preparation of a wide variety of both organic and organometallic reagents to enzyme-catalyzed reactions and the preparation of novel inorganic materials. In many cases, ultrasound was the method of last resort and has provided solutions to problems in "difficult" areas where standard techniques have been tried and failed. Hence, it is reasonable to speculate that there are many more conventional situations where ultrasound could be used effectively.

Many authors have commented on the high yields and purity of products isolated from sonochemical reactions, often in cases where other techniques have produced complex mixtures. Another factor in their favour is the reduction in the length of time required and there are many examples where conventional reactions can be brought to completion within a matter of minutes rather than hours.

We hope that this monograph provides a critical introduction for those intrigued by the synthetic potential of this technique.

We would like to thank Rod Bates, Howard Broughton, Sarah Houlton and Michael McHugh for their help in proof-reading the manuscript and for their helpful suggestions.

London, July 1989   Steven Ley and Caroline Low

Table of Contents

1 Introduction

Electromagnetic radiation whose wavelength is marginally shorter than the upper limit of perception for the human eye is termed "ultraviolet". By analogy, the term "ultrasound" is used to describe sound waves whose frequencies lie within the range 20 to 10000 kHz.

Ultrasound has become familiar to most people in connection with fields as diverse as medical imaging, non-destructive testing of materials, underwater ranging (depth gauges and SONAR imaging), and welding of thermoplastics. Ultrasonic pulse/echo techniques are used to locate mineral and oil deposits and dentists employ ultrasound to both clean and drill teeth. As a result, a wide variety of instrumentation is already available, ranging from the microtip probes used by biochemists for the disruption of cells to the enormous baths and reactors used for cleaning, degassing, or dispersion of pigments and solids in the manufacture of paints and foodstuffs on an industrial scale.

The effects of ultrasound on chemical [1] and biological [2] systems has been under investigation for over 60 years and yet its use as a tool for the synthetic chemist scarcely predates 1980. However, the widespread availability of cheap and reliable ultrasound generators, in the shape of laboratory cleaning baths, has prompted a resurgence of interest in its effects. In this period of time numerous examples of the profound influence of ultrasound on organic and organometallic reactions have appeared, most of which did not require the use of specialist instrumentation. Low amplitude, i.e. high frequency, ultrasound is also used as an analytical tool to assess the conformation of large biomolecules in solution, or for remote sensing in flow systems. This derives from the dependency of the velocity of sound on the medium through which it passes. Hence, changes in the nature of the reaction mixture, for example formation of a product, will result in a change in the emission/reception time of an ultrasonic pulse.

However, the purpose of this review is to provide a critical introduction for those intrigued by the *synthetic* potential of ultrasound and this is reflected in the examples discussed. In addition, a brief overview of the physics of acoustic cavitation is included in order to explain the origin of sonochemical reactivity and the means by which ultrasound can be exploited to its full potential.

2 The Physical Basis of Sonochemistry

2.1 The Origin of Sonochemical Reactivity

The precise details of the way in which ultrasound acts to produce chemical reaction are not known. However, ultrasonic cavitation is well established as the originator of this phenomenon. This review does not attempt to cover all the literature that has been generated on this subject, but rather to present a qualitative overview in an attempt to clarify this subject which tends to be regarded as something of a black art! For a more detailed report, interested readers are referred to Suslick's recently published review which deals comprehensively with this aspect of sonochemistry [3, 4].

The chemical effects of ultrasound do not arise from any direct input of sonic energy to species on a molecular level. A quick examination of the physical data for water can be used to illustrate this. The velocity of sound in water is 1500 ms^{-1} and the frequency range that is designated "ultrasound" is between 20 and 10,000 KHz. Hence, the wavelength of the vibrations generated is in the region 7.5 to 0.015 cm. That is, the direct energy input is not high enough to produce chemical reaction. Rather, direct comparison with the electromagnetic spectrum shows that it corresponds to the energy associated with long wave radio broadcasts (Fig. 1).

The marked effects of ultrasound actually arise from the way in which sound is propagated through media. Longitudinal vibration of molecules in a liquid generates a series of compressions and rarefactions, that is areas of high and low local pressure. If the solvent molecules are torn apart with sufficient force during a rarefaction, cavities are formed at the points where the pressure in the liquid drops well below its vapour pressure. This creates

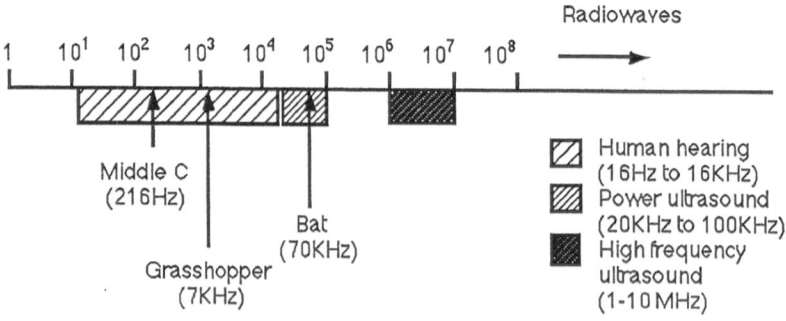

Fig. 1. The sonic spectrum

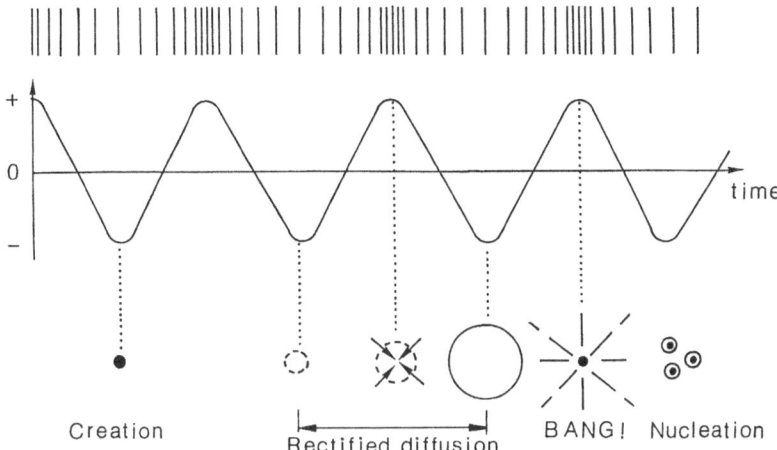

Fig. 2. Schematic representation of the lifetime of a transient cavitation bubble

a series of gas and vapour filled bubbles for which a variety of fates can be envisaged. Very small bubbles will simply redissolve as fast as they are formed. Large bubbles have long lifetimes with respect to the length of the acoustic cycle. They simply oscillate radially with the sound wave and are not responsible for the chemical effects of ultrasound. This behaviour is termed "stable cavitation".

The process by which chemical reaction occurs is defined by the behaviour of the small bubbles. Their lifetime is only a few acoustic cycles long, during which they expand to 2 to 3 times their initial size — fed by vapour pumped into the bubble as its surface expands and contracts with the changing pressure in the liquid. The bubbles then gravitate towards the pressure antinodes and collapse violently during the next compression half cycle (Fig. 2). The local pressures and temperatures generated by this collapse are enormous. This process is termed "transient cavitation" and chemical effects observed arise directly from this phenomenon. In practice, the distinction between "stable" and "transient" cavitation is not always clear cut and several workers have now presented evidence that suggests that there is a degree of interchange between the two categories.

At this point it should be stressed that *no* reaction occurs in the absence of cavitation and a number of published examples contain parallel experiments which show that vigorous stirring or mechanical agitation fails to produce such marked increases in the rate of reaction [5, 6, 7]. That is, this is not primarily a mixing effect.

Recent work by Lauterborn et al. [8] on high speed holography has produced pictures of the collapse of bubbles and the shock waves generated (see plates 1 and 2) by sonolysis of water.

The values quoted for the pressure (105 kPa) and temperatures (1000 to 3000 K) at the point of collapse have been arrived at by calculation in a number of detailed studies of bubble dynamics [9] and studies of sonolumin-

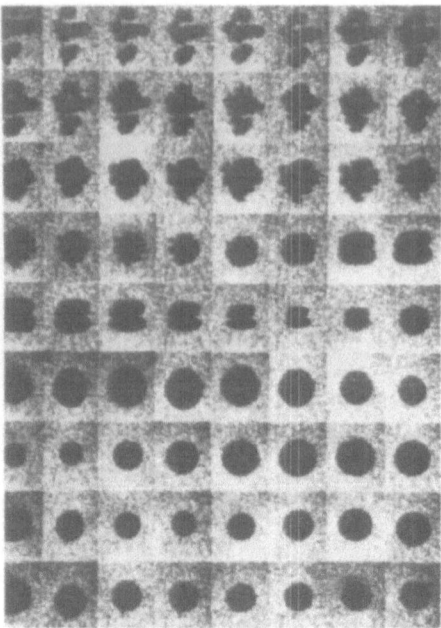

Plate 1. Spherical oscilations and decay of a single spherical bubble set into motion by a sound field of increasing amplitude at about 7 kHz. The holographic framing rate is 66.7 kHz and the frame size is 2.4×2.0 mm

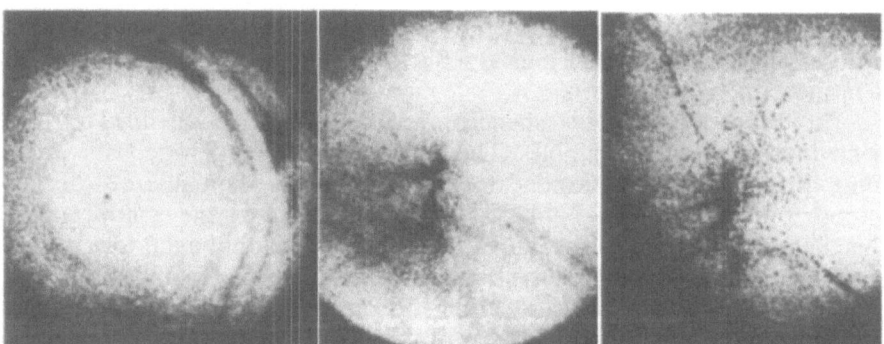

Plate 2. Collapse of a bubble field with shock wave emission. The bubble field has been produced inside a cylindrical transducer at 18.7 kHz. The holographic framing rate is 14 kHz

escence [10, 11] — another phenomenon associated with cavitation which will not be discussed here. The most recent studies, carried out in heptane/decane mixtures, gave a figure of 5200 K for the effective temperature in the bubble [12]. But the data, obtained using techniques developed in shock tube experiments, also suggested that the reaction did not occur entirely in the vapour phase. The authors proposed that the thin liquid "shell" surrounding the cavity was the second site of reaction and calculated that the effective temperature in this region was about 1900 K (Fig. 3). Hence, the mechanism by which reaction occurs is primarily thermal. However, despite the magnitude of the figures quoted, the macroscopic temperature change

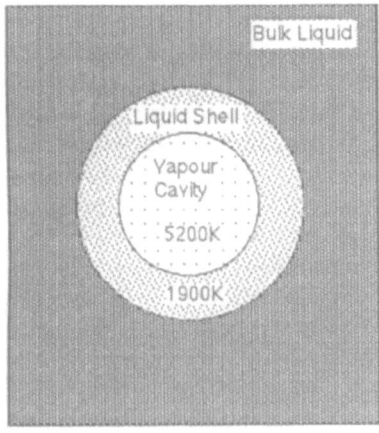

Fig. 3. Schematic diagram showing the calculated values for the effective temperatures immediately prior to collapse of a cavitation bubble in heptane/decane under an argon atmosphere

during the course of a reaction run in a cleaning bath is typically only 10 to 15 °C — most of which occurs within the first few minutes of sonication. In a series of experiments on the saponification of a number of aryl esters, Moon et al. [5] showed that mimicking the macroscopic heating effects of ultrasound by immersing the reaction vessel in an oil bath at 100 °C for 10 min only gave 13% of the product obtained by sonolysis over the same period of time. In fact, 10 min sonolysis produced yields equivalent to 90 min at reflux. This has clear implications for reactions involving sensitive reagents or products and ultrasound is clearly useful in cases where competing thermal decomposition under equivalent thermal conditions reduces isolated yields.

The theory of "hot-spot" pyrolysis on collapse of the cavitation bubbles, which was first proposed in 1950 [13], has become generally accepted. However, a number of other theories have also been proposed to account for sonochemical behaviour. Historically, the first of these arose from observations that water emits light when exposed to ultrasound [14]. This process, termed "sonoluminescence", was proposed to occur via a mechanism similar to that which generates lightning. Charge separation in the atmosphere tears apart drops of water with a build up in potential. The rapid discharge of this energy results in a flash of lightning. Hence, it was suggested that the cavitation bubbles were created as lenticular rather than spherical cavities in the first instance; "enormous electric stresses" within the bubble were invoked. These were thought to lead to the generation of moving charges in the surface layer of the bubble. In the limiting case where charge separation is complete, i.e. the positive charges are distributed diametrically opposite the negative charges, the electric field strength across the bubble was calculated to be 600 V cm^{-1} [15]. That is, reactions were thought to be primarily electrochemical in nature.

This theory dominated sonochemistry until the 1970s gaining credence from observations that sonolysis of water produced similar species to those of electrical discharge or radiolysis. More importantly, it was also responsible for the premise that organic liquids could not support high intensity cavitation. This was suggested on the basis that the dielectric constants of organic solvents are considerably lower than water. Dogmatic statement of this theory must be held responsible for the lack of recognition accorded to the first observations of chemical reaction in organic solvents which had begun to appear in the early 1950s. Schulz and Henglein reported that sonication of diphenylpicrylhydracyl, a well known radical trap, in methanol resulted in decolourisation of the solution [16]. Further work showed that chloroform decomposed releasing hydrogen chloride [17], although it was only recently that the presence of accompanying free radicals and carbenes was established [18]. It should be stressed that great care was taken to ensure that water was rigorously excluded from the system, with respect to the dogma above.

Despite this, it was not until 1980 that a detailed study of cavitation intensity in a range of organic liquids was published [19] (Table 1), and almost all the literature concerning reactions in non-aqueous solvents postdates this. Unfortunately, these results are only of value in a qualitative sense, since they were obtained using a white noise ultrasound generator, i.e. a cleaning bath. Hence, for reasons which will become clear later, it is not surprising that they show no clear correlation with any of the solvent's chemical or physical properties.

Suslick et al. [20] later demonstrated the link between the intensity of caviation in an organic liquid and its vapour pressure. Sonolysis of $Fe(CO)_5$

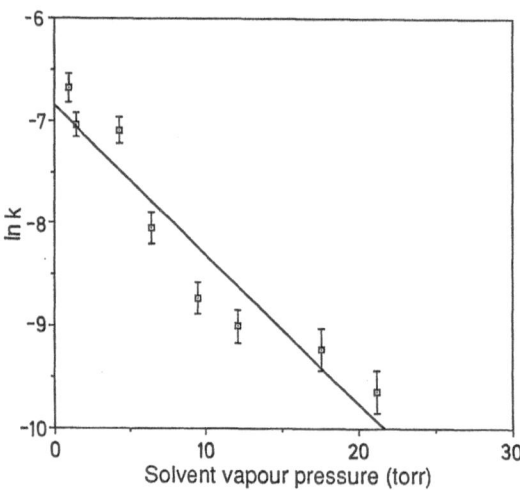

Fig. 4. Variation of the rate of decomposition of $Fe(CO)_5$ with the solvent vapour pressure at 25 °C, under argon, for a series of hydrocarbon solvents (in order of increasing solvent pressure): decalin, decane, nonane, 0.22 mole fraction octane in nonane, 0.52 mole fraction octane in nonane, octane, 0.11 mole fraction heptane in octane. Vapour pressures were calculated assuming ideal solution behaviour at an acoustic intensity of 80 W cm^{-2}

Table 1. Comparison of ultrasonic cavitation intensity in various liquids [19]

Liquid	bp/°C	Maximum cavitation intensity for a $\lambda/2$ liquid column (46 kHz) [%]	Temperature at which cavitation reaches maximum intensity/°C
1. Water	100	100	35
2. Styrene	146	74	37
3. Toluene	111	71	29
4. Tetralin	207	70	55
5. Cyclohexanone	155	70	36
6. Morpholine	128	65	50
7. Xylene	137	64	26
8. Ethylene glycol	197	61	93
9. Cyclopentanol	141	59	49
10. Trichloroethylene	87	58	20
11. Glycerine	290	57	85
12. n-Amyl acetate	149	57	18
13. Tetrachloroethylene	121	56	42
14. n-Butyl acetate	126	56	21
15. Pyrrole	130	55	40
16. Methanol	65	52	19
17. Chloroform	61	50	−3
18. n-Amyl alcohol	137	47	23
19. Ethanol	78	46	21
20. Ethyl acetate	77	45	9
21. Acetone	56	44	−36
22. n-Butanol	118	43	32
23. Benzene	80	43	19
24. n-Propanol	97	42	27
25. 1,1,1-Trichloroethane	74	41	18
26. Dichloromethane	40	38	−40
27. Methyl acetate	57	38	−32
28. Naphtha	242	38	35
29. Isopropanol	82	38	16
30. Formic acid (85%)	101	37	30
31. Tri-n-butylamine	214	37	31
32. Tetrachloromethane	77	35	8
33. Cyclohexanol	160	23	37
34. Propanoic acid	141	22	32
35. Triethylamine	89	29	1
36. Freon 113	48	15	−20
37. Freon 114B2	47	6	8
38. Acetic acid	118	6	48

results in thermal dissociation of the metal carbonyl. The products of this reaction are $Fe_3(CO)_{12}$ and finely divided iron. Careful investigation has shown that the rate of this reaction is inversely proportional to the vapour pressure of the surrounding solvent (Fig. 4). These findings are justifiable in terms of the "hot spot" mechanism; that is, the adiabatic heating effects of rapid compression reduce as the vapour recondenses. Secondly, conduction of heat to the surroundings reduces the local heating effect still further. The overall effect is that increasing the vapour pressure reduces the intensity of cavitational collapse, the maximum temperature attainable, and consequently the rate of the ongoing reaction.

Hence, ultrasonic reactions have the curious property that reaction rates decrease as the ambient temperature increases. The electrical discharge theory, which still had its supporters in the mid-1970s, has finally been rejected as being inconsistent with experimentally determined results [15, 19]. Another theory, proposed to account for the degradation of large polymers in ultrasonic fields, suggests that direct mechanical cleavage of bonds may be occurring as a result of the intense shock waves generated by transient cavitation, or indeed, the direct effect of the accelerations produced by the sound field that have been estimated to be as high as 105 g at 500 kHz [21]. However, interaction of the polymer with reactive species produced as a result of solvent breakdown could also account for this.

The most recent proposal was put forward as a result of studies carried out using binary water/ethanol mixtures [4c]. In considering the solvolysis of t-butyl chloride, Mason and co-workers surmised that the rate of reaction must be dependent of the degree of solvation in the transition state prior to ionisation. The reaction was monitored by conductimetric measurement of HCl (Scheme 1).

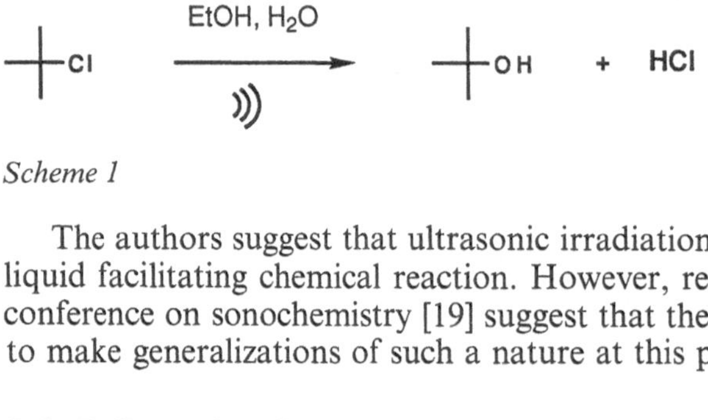

Scheme 1

The authors suggest that ultrasonic irradiation alters the structure of the liquid facilitating chemical reaction. However, results presented at a recent conference on sonochemistry [19] suggest that the system is too complicated to make generalizations of such a nature at this point in time.

2.2 Influencing Sonochemical Reactivity: The Physical Perspectives

2.2.1 Introduction

The following chapters contain numerous references dealing with the profound effects of exposing chemical reagents to ultrasound. However, the majority of reactions cited were carried out in laboratory cleaning baths where the scope for changing the reaction parameters in extremely limited. Hence, it is probably true to say that the majority of these have not been fully optimised.

For example, with regard to the "hot-spot" theory outlined above, it would clearly be useful to understand the effects of changing the solvent, or the ambient temperature and at which the reaction was carried out. Furthermore, the design of ultrasonic probe systems allows for ready variation of the power input and occasionally variation of the frequency of the output. Hence, there are a number of factors that must be born in mind when setting up a viable system. For this reason, the following section is devoted to a discussion of the effects of extrinsic variables on the sonochemical process.

2.2.2 Frequency

The credibility of the "hot spot" theory is reinforced by its ability to account for the effects of extrinsic variables on the sonochemical process. Nevertheless, the frequency of ultrasound applied is surprisingly irrelevant to the course of the reaction. Cleaning baths produce a range of frequencies which often vary from day to day, or even during the course of a reaction, and yet this has no discernable effect on the sonochemistry observed.

Experiments have shown that aqueous sonochemistry is unchanged over the frequency range in which cavitation occurs i.e. 10 Hz to 10 MHz [22]. Since there is no direct coupling of the sound field with species on a molecular level, changing the frequency of the sound input simply alters the resonant size of the cavitation bubble. The effect of this over the range of interest is negligible. It should, however, be noted that although there is both an upper and a lower limit to the frequencies at which cavitation will occur, the band of frequencies used for sonochemistry lies well within these limits.

It has also been shown that ten times more power is required to make water cavitate at 400 kHz than at 10 kHz. This effect is due to the increased power losses which occur as the rate of molecular motion within the liquid increases. Hence, there is no advantage to be gained from using frequencies higher than those which can be obtained using a simple cleaning bath.

2.2.3 Power Input

Conversely, changing the power input to the transducer alters the volume of liquid which can be forced to cavitate and dramatically affects the observed sonochemical rate. Hence, a number of reactions which give low or erratic yields in cleaning baths, produce high yields when a probe transducer is used, as a result of the ability to control the acoustic intensity input. There is a school of thought that an optimal value for the energy input exists. However, Luche has reported that the rate of the Barbier reaction between an alkyllithium and an aldehyde increases continuously as the voltage input to the transducer is increased from 60 to 160 V (Fig. 5) [23].

Luche stresses that the results of these experiments do not represent

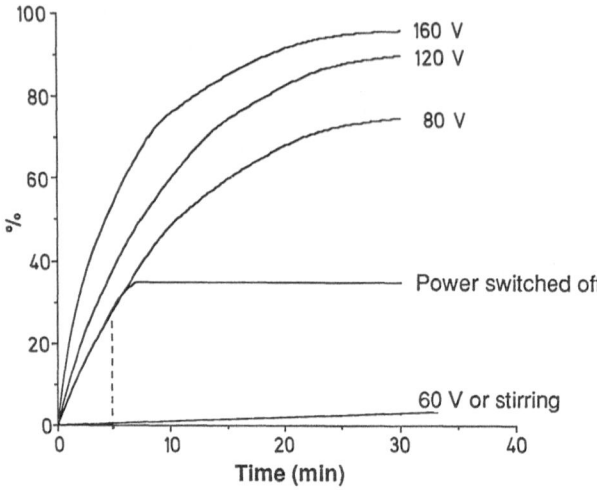

Fig. 5. The effects of increasing power input on the rate of the reaction between an alkyllithium and benzaldehyde

true kinetic data, as the temperature could not be stabilised, especially during the first few minutes. However, the overall accelerating effect of sonication is obvious from Fig. 5. Furthermore, Luche quotes the power input in terms of the electrical input to the transducer. However, the proportion of energy used to effect chemical change cannot be deduced from this simplistic view.

Increasing the acoustic intensity also extends the range of bubble sizes that will undergo transient cavitation. However, experimental observations show that there is an upper limit beyond which no additional rate increase is seen. Explanations put forward suggest that penetration of sound into the body of the liquid is hindered if cavitation is so intense that the radiating surface becomes shrouded in a layer of bubbles. Conversely, it may simply be that bubble growth becomes so rapid that the boundary for transient cavitation is exceeded before the next compression half-cycle [24].

2.2.4 Bulk Temperature

Having discussed the effects of solvent vapour pressure on the implosion of cavitation bubbles, it might be expected that lowering the ambient temperature of the medium would lead to increased rates of reaction. However, several workers have shown that an optimal temperature exists [23, 25]. This can be explained if one assumes that the number of cavitation nucleii present will increase with increasing temperature to the point at which the increase in vapour pressure dominates the reactivity of the system. For example, Rosenberg examined the effects of ultrasound on the erosion of aluminium [25f]. Examining the system between $-10\,°C$ and $90\,°C$ showed that the degree of erosion peaked at $50\,°C$. Somewhat surprisingly, this result was shown to be consistent in a variety of liquids.

2.2.5 Pressure

The effects of increasing the static pressure are unclear. It could be reasoned that if the total pressure on the system were greater, one would expect the intensity of cavitational collapse to increase. However, reports differ wildly and the views expressed can truly be said to cover every possibility [26]! The situation is further complicated by consideration of the events which lead to nucleation of the bubbles.

At the present, the generally accepted mechanism for nucleation of bubbles (Fig. 6) suggests that gas trapped in small angle crevices of particulate contaminants expands and contracts with the acoustic cycle. Free air bubbles would not be expected to act as nucleation sites on the basis that they are inherently unstable under these conditions and would be expected to dissolve as a result of surface tension. As the bubble volume grows two possibilities arise: on the one hand, small gas bubbles may be released into the surrounding liquid; and on the other, implosive collapse of the bubble will release a stream of microcavities at which nucleation can occur.

This theory was proposed in the light of observations that ultrafiltration of the solvent increases the cavitation threshold, as do both pressurization and evacuation of the system. Hence, nucleation is supressed by the flooding of the crevices.

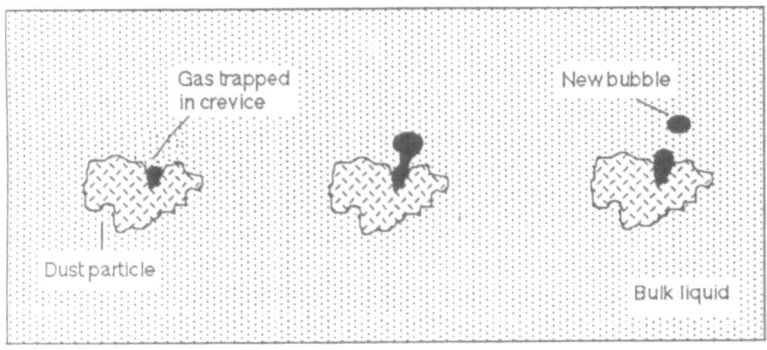

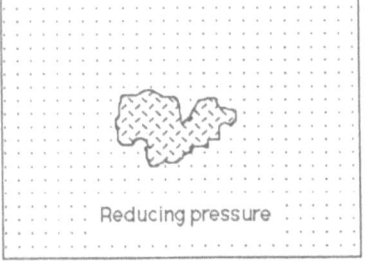

Fig. 6. Schematic diagram showing the generally accepted mechanism for the nucleation of cavitation bubbles

As previously noted, experimental results obtained on increasing the hydrostatic pressure are not in full agreement [26a, 27]. However, this is probably a reflection of the difficulties encountered in maintaining constant temperature in a pressure vessel under ultrasonic irradiation. Recent work on the oxidation of cyclohexane confirms the predictions of the "Hot-spot" theory and the authors report an increase in rate to a maximum value after which it begins to fall. Thus it was concluded that the collapse of the cavitation bubbles was indeed more efficient. However, formation of the cavities becomes simultaneously more difficult [28].

2.2.6 Ambient Gas

Experiments have also shown that sonolysis reactions occur faster in the presence of monoatomic gases than those carried out in the presence of diatomic gases. In fact, the maximum temperature reached during cavitation is strongly dependent on the polytropic ratio of the ambient gas ($\gamma = Cp/Cv$), its thermal conductivity and its solubility.

The former is a measure of the heat released on adiabatic compression of the gas and the latter has a bearing on the number of cavitation nucleii available. Furthermore, it should be noted that H_2, N_2, O_2 and CO_2 are not inert during cavitation and will undergo a variety of redox and radical reactions [29] and addition of a gas with a high solubility e.g. CO_2 was shown to arrest formation of the "dark-coloured compounds" observed on sonolysis of nitrobenzene [30].

2.2.7 Choice of Solvent

The chemical reactivity of the solvent must also be taken into account. *No solvent is inert under cavitation conditions* and use of halogenated solvents should be avoided. This stems from one of the earliest reports of the effects of ultrasound on an organic solvent when Henglein showed that hydrogen chloride was released from chloroform [17]. The other products of its decomposition were later shown to be a variety of free radical and carbene species [18]. In addition, sonolysis of carbon tetrachloride produces elemental chlorine and acetonitrile releases hydrogen, nitrogen and methane on exposure to ultrasound [31]. Aromatic hydrocarbons darken as polymerization occurs on long exposure to ultrasound [30] and even linear alkanes undergo cracking under high intensity ultrasound [32].

As previously discussed, the intensity of cavitational collapse is dependent on the vapour pressure of the solvent and use of involatile solvents is recommended [3], although there are a large number of examples where ethereal solvents have been used to good effect. Other liquid properties such as viscosity and surface tension may have some bearing on the threshold at which cavitation will occur, but these are of minor concern.

Aqueous sonochemistry is dominated by reactions of OH and H radicals as a consequence of the high vapour pressure of water relative to any organic or inorganic reagents present and despite the enormous amount of attention

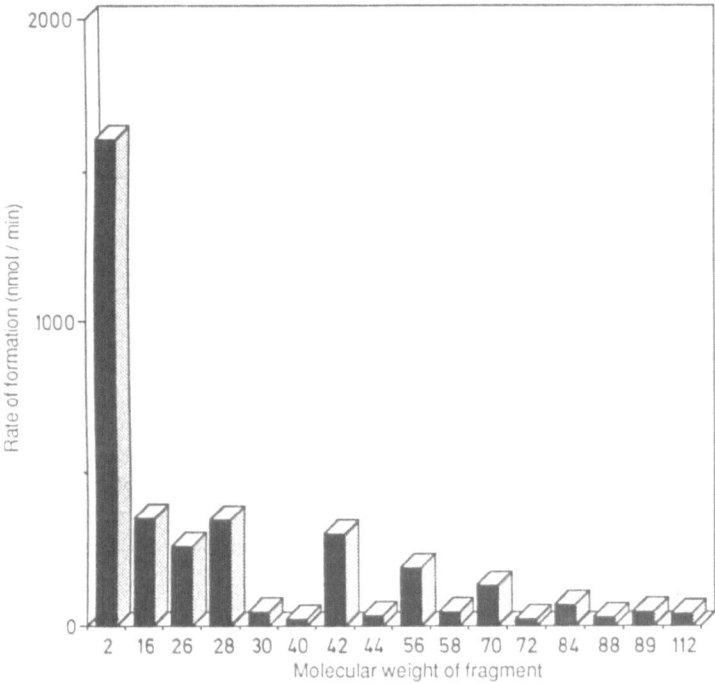

Fig. 7. The distribution of products formed on sonolysis of *n*-decane under argon at an acoustic intensity of 100 W cm^{-2} and 25 °C

it has received [33] the number of synthetically useful reactions generated is extremely small (See Chapter 3 for details).

With careful consideration of these factors, it is clearly possible to 'fine tune' systems to maximise the efficiency of the cavitation process and an understanding of the effects of these parameters on sonochemical processes is clearly of help when attempting to optimize the rates and yields of these reactions.

2.3 The Effects of Ultrasound on Two-phase Systems

As previously explained, the effects of ultrasound on homogeneous systems are dominated by the enormous changes in pressure and temperature created at "hot spots" on implosion of the cavitation bubbles. However, in two-phase systems a number of other factors must be taken into consideration. The relative contributions of these phenomena have not been conclusively established in any one case. It appears that effects seen in liquid/liquid systems are principally due to emulsification which occurs when the shearing stresses on the liquid are greater than the interfacial surface tension. In a number of cases this enormous increase in surface area of contact completely obviates the need for use of phase transfer catalysts [5, 34] greatly simplifying the work-up procedure for such reactions.

There are also examples where the main effects of sonication arise from improved mass transport within the system. An example of this is the rate increases reported in electrolysis reactions [35]. The reason for this lies in the increase in momentum brought about as the solvent absorbs energy from the propagating sound wave. This phenomenon is independent of cavitation and is known as "acoustic streaming".

At liquid/solid interfaces the effects of cavitation are seen in the well known cleaning effects of ultrasound. Metal surfaces are eroded and soft metals like potassium can be reduced to colloidal suspensions by low intensity ultrasound [36]. These effects, coupled with those already detailed, are responsible for the enormous increases in rates of reaction that have been reported in heterogeneous systems. Such reactions form the bulk of those described in this review. Most of these can be carried out using a simple cleaning bath as the source of ultrasound. In comparison, homogeneous reactions generally require a more intense and specific input of energy such as that provided by a probe generator. Reactions occurring between species in solution and solid reagents are influenced by a greater number of factors than those which occur within the same phase. The overall rate of reaction is hence a combination of the rates at which species diffuse to and away from the surface as well as a reflection of the number of reaction sites available.

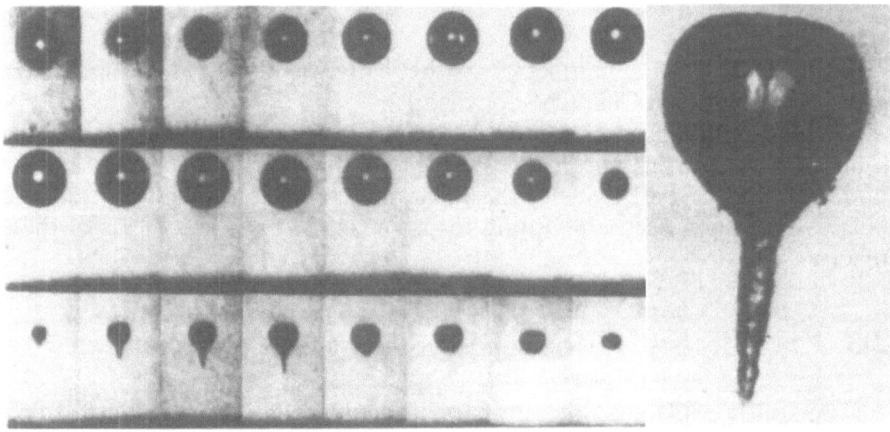

a   b

Fig. 8. a Jet formation upon collapse of a spherical cavity in the neighbourhood of a plane solid boundary in water. Maximum bubble radius 1.1 mm; distance of bubble centre from the wall 4.5 mm; framing rate 75000 frames/s. [The first frame (upper left) is doubly exposed with another cavity].
b Cavity in water during its first rebound after collapse near a solid boundary below a typical, well developed jet

Ultrasound seems to act on all these fronts. For instance, acoustic streaming improves mass transport between the bulk of the liquid and the surface. Similarly, the cleaning effects of ultrasound mean that species can be removed from the surface with equal ease increasing the effective number of sites available for reaction. Cavitation causes erosion of the surface and the intense pressures and temperatures generated on implosion of the bubbles may cause defects and deformations within the solid. Furthermore, calculations suggest that the shock waves generated are of sufficient magnitude to cause plastic deformations in malleable metals (approx. 104 atmospheres) [37]. The cumulative result is to increase the actual number of reaction sites on the surface. This is clearly shown in electron micrographs of surfaces activated in this way [4g, 38].

High speed microphotographs of bubbles collapsing at a solid boundary show that implosion of the cavity is no longer spherical, but the slight asymmetry is self reinforcing and results in elongation of the bubble to produce a whirling jet. Tip velocities have been estimated to be in excess of 100 ms^{-1} and erosion of the surface occurs at the point of contact [32] (Fig. 8).

Electron microscopy shows that crystalline metal faces show initial microscopic pitting of the surface and plastic deformation. Continued exposure results in grain boundary delineation and finally, large scale cratering. Close examination shows that sonication of a powdered solid results in a reduction in particle size over the first few minutes. For example, particles of between 60 and 90 μm diameter showed a reduction of 5 to 10 μm. However, continuing sonication had no further effect. This is to be expected if one considers that the cavities formed by ultrasonic irradiation are of the order of 10 to 100 μm in diameter and can only form jets at surfaces with at least those dimensions. Secondly, the effect of shockwave accelerations over the whole particle decreases proportionally with size and hence, small particles do not undergo further fragmentation. The corollary of this is that the available surface area of a solid such as TaS_2 increases 10 fold after 15 minutes irradiation (Plate 3). In contrast, the surface area of nickel powders with an initial particle size of 3 to 5 μm only changes from 0.48 to $0.69 \text{ m}^2 \text{ g}^{-1}$. Examination of the surface morphology after sonication shows a marked difference between the layered metal chalcogens and the metal powder. The former appears as a badly eroded pockmarked surface, whereas the latter appears to have been smoothed over. Electron micrographs of manganese and lithium (Plate 4) show similar effects [4g, 38]. The different results are probably due to the difference in ductility between the brittle inorganic solids and the malleable metals.

Luche and co-workers have undertaken a detailed investigation of the effects of ultrasound on lithium in the presence and absence of chemical reagents [38]. It could be postulated that the primary role of the ultrasound is to disperse the lithium removing small, highly activated particles from the metal. However, it has already been shown that this does not occur to any noticeable extent in THF [36], the solvent in which these investigations were carried out. Alternatively, cavitation must lead to modification of the metal

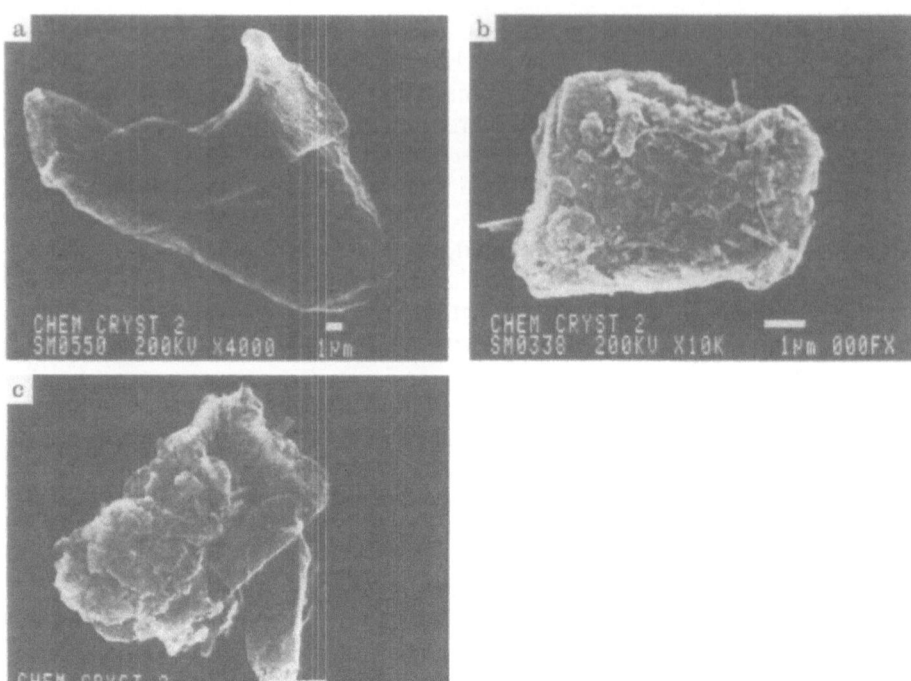

Plate 3a–c. Scanning electron micrographs showing the reduction of particle size obtained on sonication of TaS$_2$: **a** TaS$_2$ before ultrasonic irradiation, **b** after 10 min irradiation, and **c** 120 min irradiation (Note change in scale)

surface. Scanning electron micrographs of lithium surfaces under chemical attack show the formation of shallow pits which, by analogy with work done on the initiation of Grignard reagents, are reasoned to act as initiation sites and the initial rate of the reaction is a direct function of the density of these sites. Sonication for 15 minutes produces a vast increase in the available surface area, although the overall effect is one of an irregular disorganised surface (Plate 4b). By contrast, sonication of the metal in the presence of chemical reagents reinforces the effects of chemical attack alone. However, the density of craters formed is many times higher than that obtained in its absence (Plate 4c).

Hence, sonication produces highly active, unpassivated surfaces and although the relative contributions of the effects outlined have not been fully established; the beneficial effects of ultrasound in such systems are clearly reflected in the large number of literature examples where it has been used to good effect.

Plate 4a–c. Scanning electron micrographs showing **a** the original state of the lithium ▶ (magnification 500), **b** the effects of sonication in THF at room temperature, in the absence of chemical reagents for 15 minutes at 860 V (magnification 50), and **c** after 5 min combined chemical attack and sonication at 860 V and 20 °C (magnification 50)

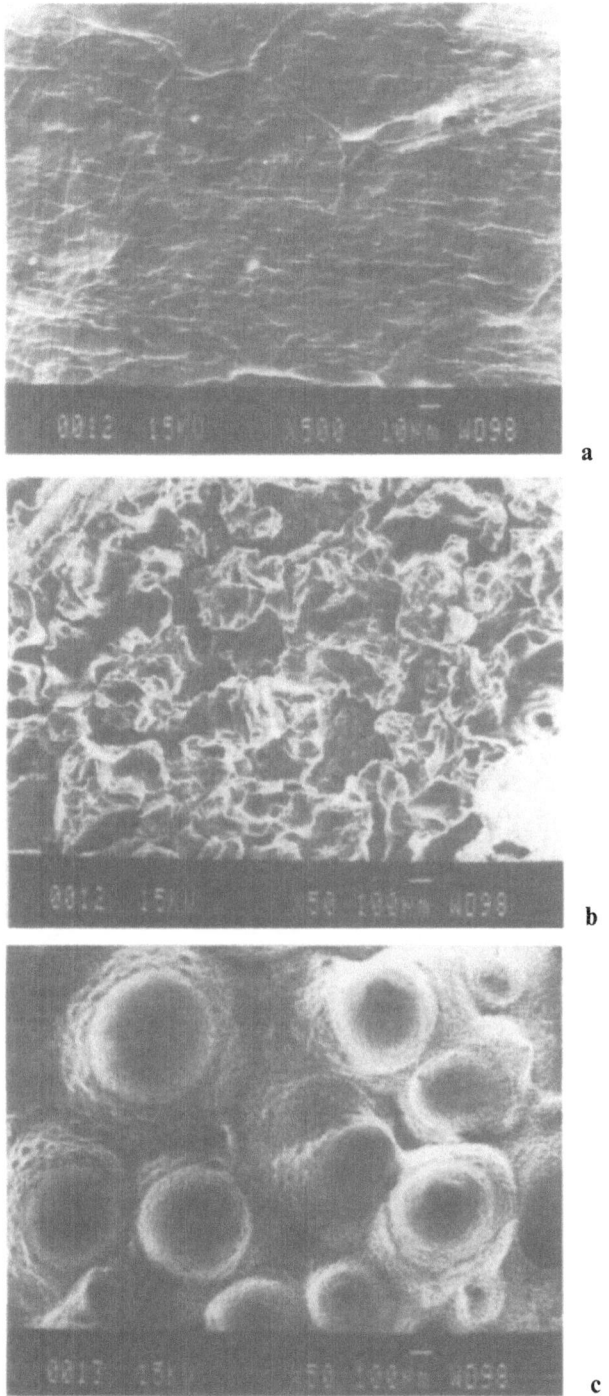

a

b

c

2.4 Generation of Ultrasound

It is probably true to say that most workers have access to the small ultrasound baths commonly used for cleaning apparatus or emulsification of samples. Their widespread availability is responsible for the burgeoning number of reports concerned with the effects of ultrasound on heterogeneous systems. In most cases, the reagents are simply mixed together and the reaction flask suspended in the cleaning bath in the position where maximum agitation of the flask contents is observed (Fig. 9). The simplicity of this procedure has resulted in a number of fortuitous discoveries by experimenters previously unconnected with this field. However, it should be noted that there are several potential drawbacks to the use of cleaning baths which limit their general applicability; and this restriction has led to the development of ultrasound "probes" by modification of the ultrasonic cell disruptors originally developed for use by biochemists.

In all cases, the ultrasound is generated by the expansion and contraction of a piezoelectric device in the presence of a fluctuating electric field, such as that produced by a high frequency AC voltage. Most cleaning baths have one or two such devices situated at the base of a metal bath. The ultrasound is conducted to the site of action by propagation through a volume of water. The disadvantages of this configuration lie in the lack of control that can be exercised over a number of important parameters. For instance, the acoustic

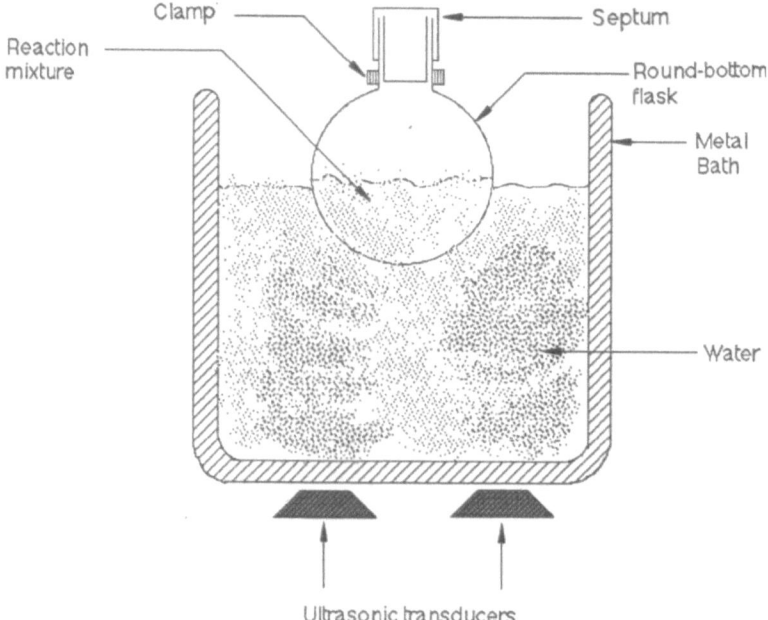

Fig. 9. Experimental set-up for reaction carried out in a standard laboratory cleaning bath

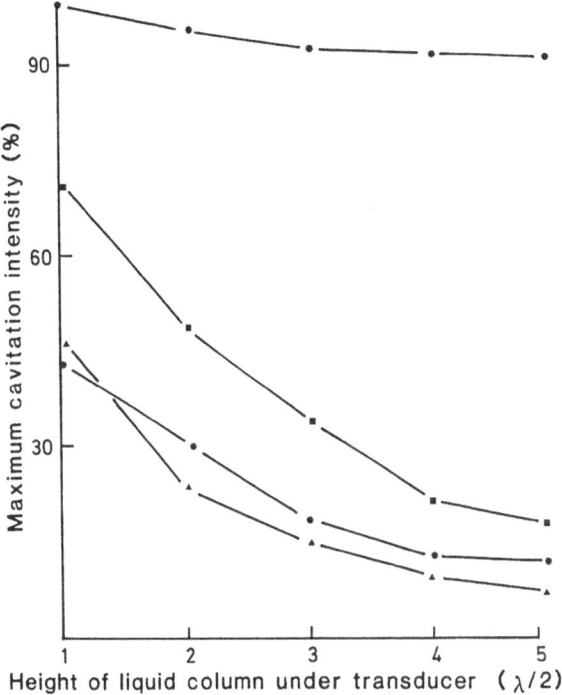

Fig. 10. Maximum cavitation intensities (at 46 kHz) of four liquids as a function of the height of the liquid column under the transducer: ● — water; ■ — toluene; * — benzene; ▲ — ethanol

intensity and frequency of ultrasound generated by cleaning baths can vary considerably from manufacturer to manufacturer and, more importantly, over the lifetime of a particular bath. In addition, the positioning of the flask and heights of liquid in both the bath and the reaction vessel become critical to the reproducibility of the reaction (Fig. 10) by virtue of the observation that cavitation will not occur at the nodes of standing waves set up in the bath [33]. Temperature control in such systems is virtually impossible and the ambient temperature is generally about 35 °C. This is an important consideration in the view of the inverse relationship between the vapour pressure of a liquid and the maximum intensity of cavitation attainable. Some authors have reported that cooling could also be achieved by replacing the water in the bath with detergent/ice, or by passing coolant through a copper coil suspended in, but not touching the walls of the bath. However, attenuation of the ultrasound as it passes through the water in the bath means that the acoustic intensity is only marginally higher than the lower limit necessary for cavitation and relies on the presence of solid "impurities" within the liquid to lower its tensile strength before cavitation can occur.

This limitation has led to the development of a number of systems that can be used to induce cavitation in homogeneous reaction mixtures. The cup-horn reactor was originally designed for use by biochemists as a cell-disruptor (Fig. 11). Its advantages are that it can deliver greater acoustic intensities and is potentially easier to thermostat. Its disadvantages are similar to those of the cleaning bath in that it is very sensitive to liquid levels and the

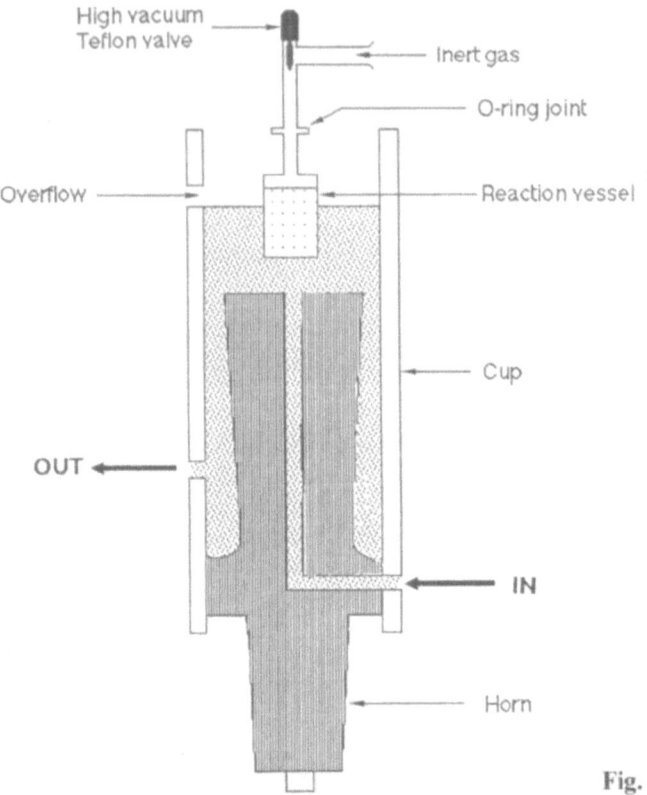

High vacuum Teflon valve

Inert gas

O-ring joint

Overflow

Reaction vessel

Cup

OUT

IN

Horn

Fig. 11. A cup-horn reactor

shape of the reaction vessel. Furthermore, there is the added limitation that the reaction vessel cannot be more than about 5 cm in diameter, thus constraining it to use in small scale reactions. Suslick has given full consideration to the merits of these designs and clearly favours ise of the direct immersion probe on the grounds that it allows access to a wide range of acoustic intensities with a high degree of reproducibility [3]. Furthermore, temperature control is readily achieved by immersion of the reaction vessel in a cooling bath and the availability of commercial flow cells allows processing of multi-litre quantities of reagents (Fig. 12). Fears that erosion of metal from the surface of the probe would lead to contamination of reaction mixtures were dispelled in a careful study published some years ago [41] and corrosion of the horn by reaction media seems unlikely in view of the low reactivity and high tensile strength of titanium.

Careful design of the reaction vessel allows reactions to be carried out under inert atmospheres (Fig. 13) or at moderate pressures (< 10 atmospheres) (Fig. 14). Other workers have proposed modifications to allow the reaction mixture to be simultaneously stirred. These include use of a cell with a small indentation at the bottom or a glass rosette cell (Fig. 15). Luche and coworkers have carried out extensive investigations into sonochemical preparation of

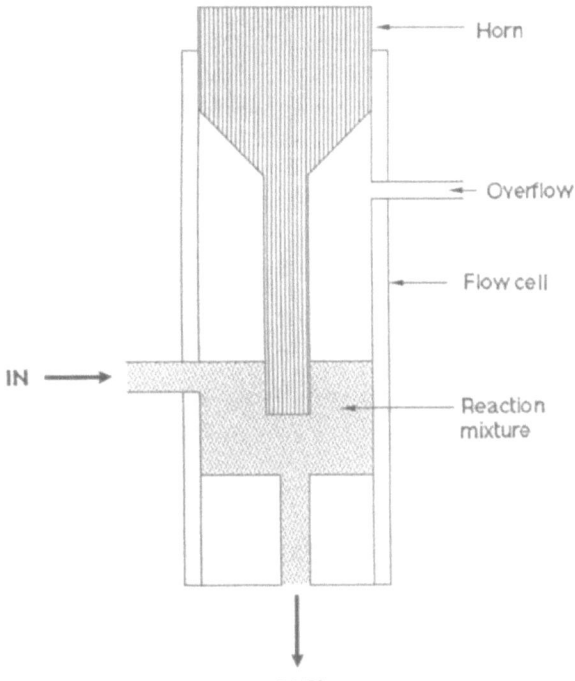

IN →

OUT

Fig. 12. Flow cell for processing multi-litre quantities of material

organolithium species and have designed a reactor containing a support on which the lithium can be placed at a fixed distance from the horn (Fig. 16). In all cases, the reaction vessels are equipped to allow reactions to be carried out under a constant stream of inert gas, a condition necessitated by the tendency of ultrasound to degas liquids.

Ultrasound is currently employed in the preparation of emulsions on an industrial scale. In this case, the reagents are pumped through a minisonic homogenizer or whistle reactor (Fig. 17). Cavitation occurs as the fluid flows across a vibrating plate and the power obtainable is limited by this factor. Most of the chemical effects observed arise from the vast increase in interfacial area rather than the ultrasonic irradiation itself. However, its advantages stem from its proven ability to process large quantities of material in this manner.

Methods for measuring ultrasonic power have been reviewed [42] but, in short, there does not seem to be a simple method for the quantitative measurement of local ultrasonic intensity when cavitation is present. Pugin has developed a number of methods for the characterisation of sound fields in a variety of reactors [43]. These were used to develop profiles of the acoustic intensity for both cleaning probes and probe systems with a view to examining the reproducibility of reactions.

A rough estimate of the acoustic power obtained from the source was obtained by measurement of the electrical power consumption. However, as

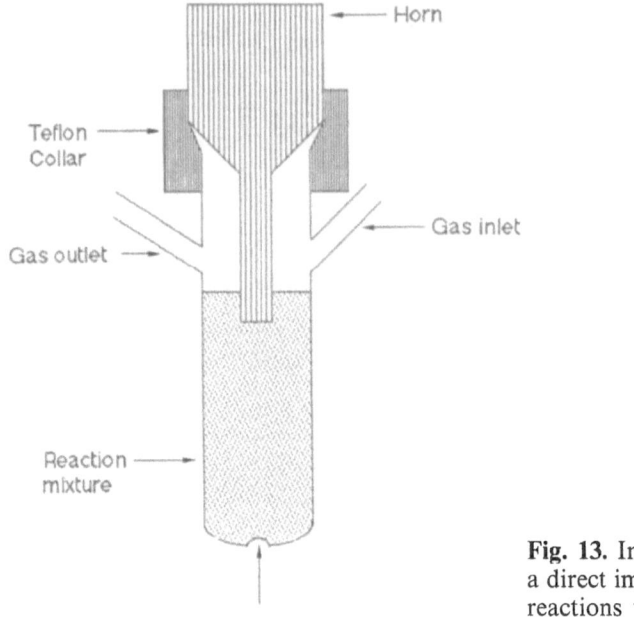

Fig. 13. Indented cell for use with a direct immersion probe allowing reactions to be carried out under an inert atmosphere

pointed out, this simplistic approach fails to take loss of intensity from cavitation, reflection or scattering into account. A chemical dosimeter was also developed in order to obtain quantitative information on the effect

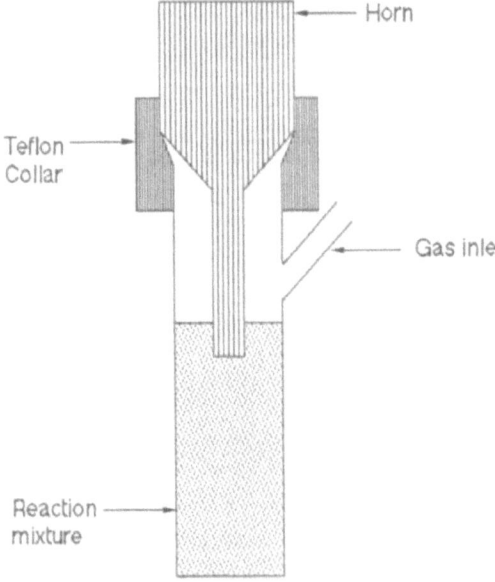

Fig. 14. Pressure vessel for use with a direct immersion probe

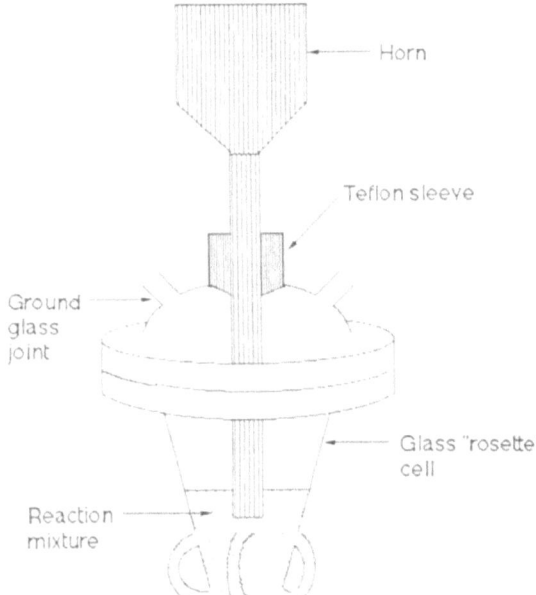

Fig. 15. Modified "rosette" cell allowing for simultaneous stirring of reaction mixture under sonication

of ultrasound on heterogeneous reactions. Pugin chose to examine the reaction between 1-bromopentane and lithium to give *n*-pentyl-lithium on

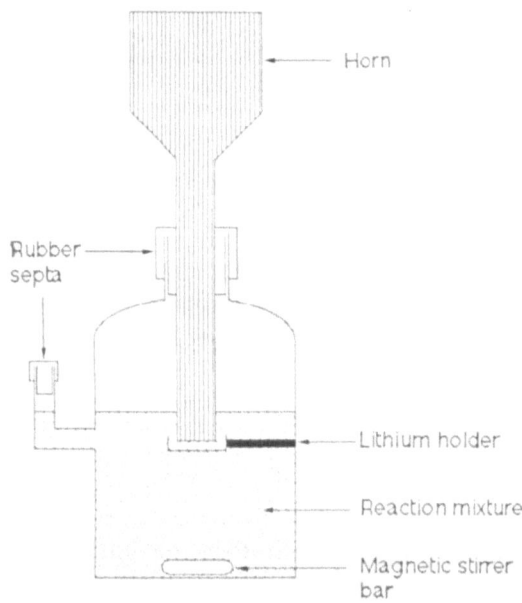

Fig. 16. Reaction vessel designed to investigate the influence of ultrasound on the preparation of organolithium reagents

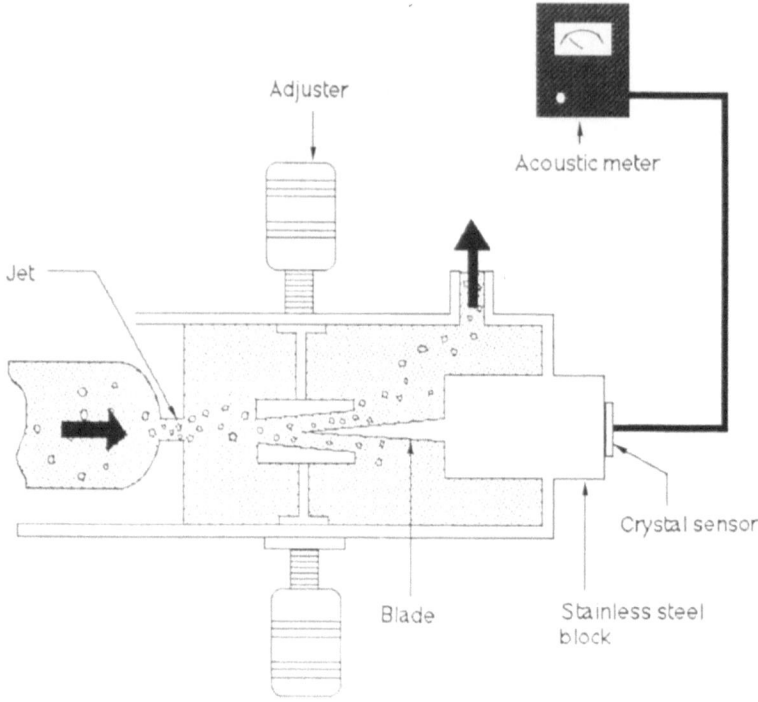

Fig. 17. Commercially available minisonic homogeniser, or "whistle" reactor

the basis that the solid reagent (in the form of lithium wire) could be fixed at any point in the ultrasonic field (Scheme 2).

Scheme 2

Of the four methods described, it transpired that the most information could be obtained by measuring the temperature of the liquid at various points in the field using a thermocouple probe. Thus, despite the lack of correlation between the power consumption of a cleaning bath and ΔT, the relationship between ΔT and the rate of reaction of n-bromopentane with lithium is approximately linear (Scheme 2), implying that the ultrasound intensity, as determined by the thermocouple probe can be directly related to the effect of ultrasound on heterogeneous reactions. This is of particular importance given the noted effects of ultrasound on such reactions. This method has the advantage that the intensity of the fields can be determined in three dimensions. Furthermore, the device employed is both simple and robust; comprising a digital thermometer equipped with a thermocouple protected by a metallic sheath (Fig. 18). The probe was sheathed in silicone rubber, which has the advantage of absorbing sound. Unfortunately,

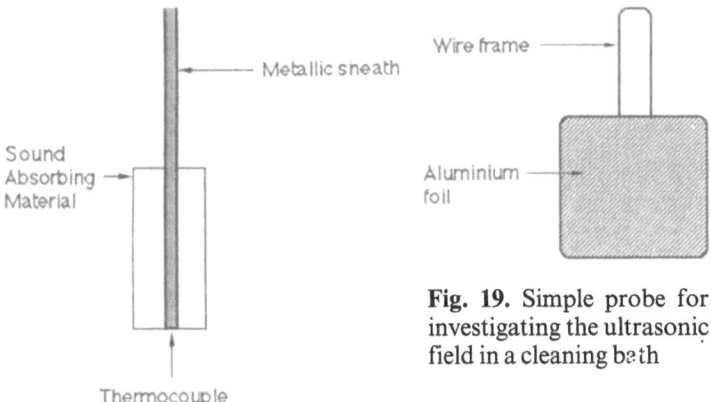

Fig. 19. Simple probe for investigating the ultrasonic field in a cleaning bath

Fig. 18. Thermocouple probe used to investigate the ultrasonic field in a cleaning bath

whilst ideally suited for measurements in water, the rubber swelled on use in organic solvents and the author finally settled on using a cork sheath although this reduced the resolution of the device. The *simplest* method described consisted of a sheet of household aluminium foil attached to a wire frame (Fig. 19). This was suspended in the bath and sonicated for a few seconds. This results in perforation of the foil. Direct comparison with measurements obtained using the thermocouple probe shows that the density of pitting on the foil corresponds well with the profile obtained using the thermo-couple probe (Fig. 20). This method can apparently be developed to provide quantitative information [44]. However, Pugin has simply used it in a qualitative sense.

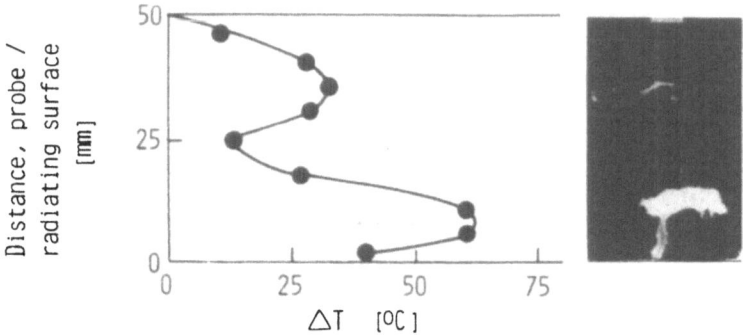

Fig. 20. Direct comparison of the output obtained from a thermocouple probe coated in silicon rubber against that obtained with the aluminium foil method in the same sound field (central axis of a Laborette 17 cleaner, water level 50 mm). The aluminium foil was sonicated for 30 s

Comparison of the effects of different sources of ultrasound are few and far between. An exception is a study of the copper-catalysed Ullman coupling of 2-iodonitrobenzene in DMF [45] (Scheme 3). The original reaction was first described by Rausch in 1961 [46] and uses a highly activated substrate, such that the thermal reaction only takes 60 h at 60 °C. It should be noted that this represents one of the lowest temperatures and highest yields reported for a classical Ullman reaction.

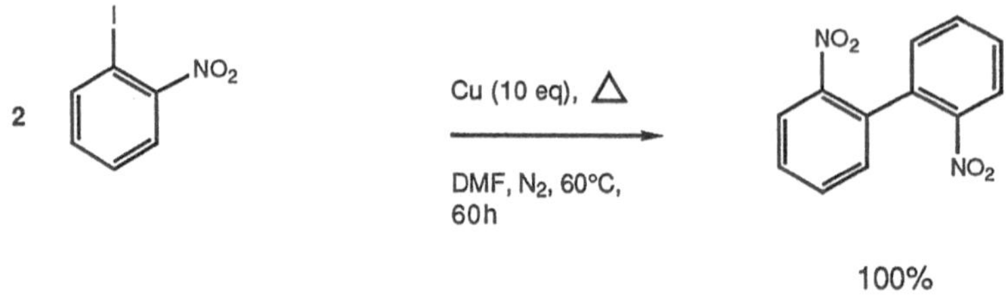

Scheme 3

Taking this as a control reaction, Lindley and co-workers investigated the effects of ultrasound on the system [45]. Initial studies using a cleaning bath operating at 40 ± 5 KHz failed to produce any marked improvement, and sonication for 2 h only gave 21 % of the biphenyl. Switching to a 1/16 inch microtip prove driven by a Heat Systems W225 sonicator (20 kHZ) operating at 60 % power showed that yields were considerably increased if the copper was activated by sonication before addition of the aryl halide. Furthermore, the reaction, which is normally carried out in the presence of a large excess of copper, was optimal at a copper to halide ratio of 4:1. However, the greatest rate enhancements were observed using a flat-tipped probe which disseminates more power into the reaction mixture. By contrast, the microtip probe provides a greater intensity of radiation although its

Table 2. Comparisons between sonochemical and photochemical apparatus

Photochemistry		Homogeneous sonochemistry	Heterogeneous sonochemistry
Source	250 W Quartz-halogen lamp	200 W cell disruptor (at 60 % power)	150 W cleaning bath
Approximate cost	$ 1800	$ 1900	$ 700
Typical rates	7 mol/min	10 mol/min	500 mol/min
Electrical Efficiency	2 mmol/kWH	5 mmol/kWH	200 mmol/kWH

effects are more localised than that of the flat-tipped probe. Hence, the reaction can be brought to completion ·within 24 h under these conditions. Interestingly, the authors declare themselves to be well aware of the link between solvent vapour pressure and cavitation intensity. Nevertheless, the

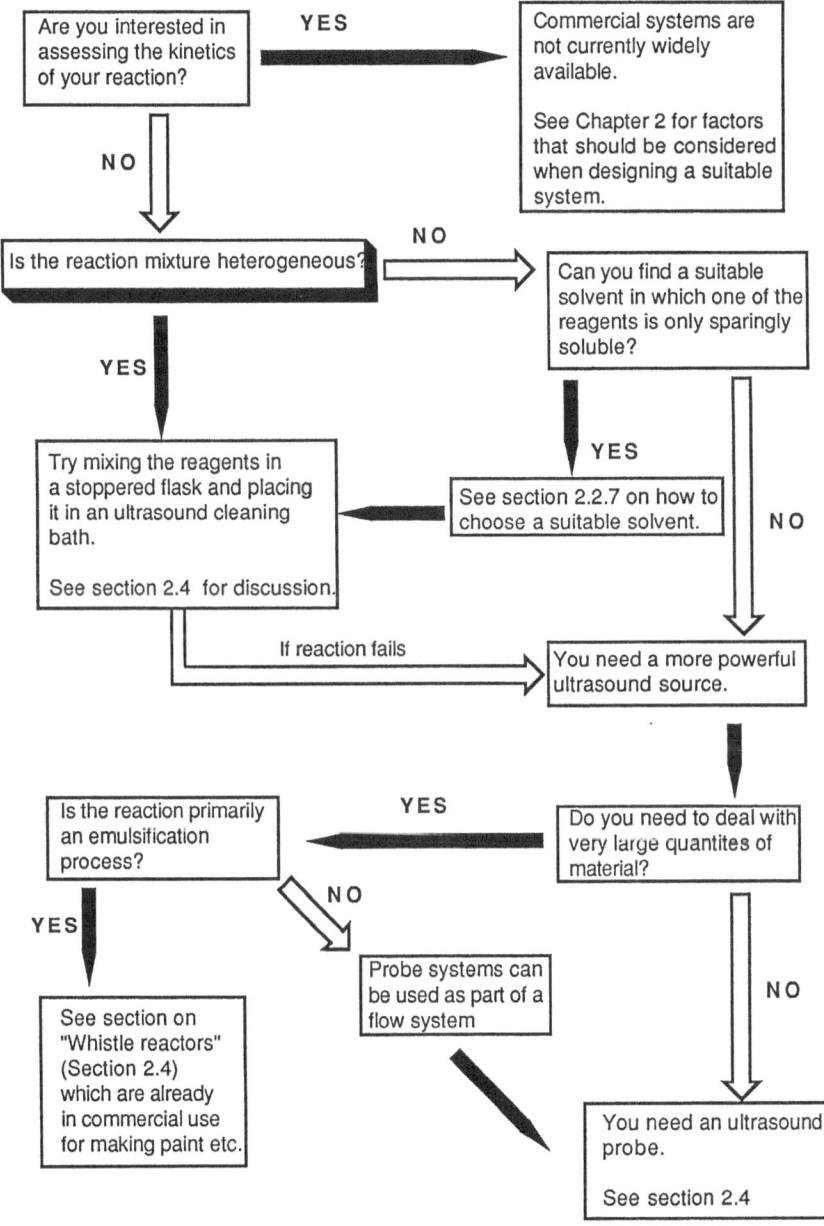

temperature of the reaction mixture was maintained above 60 °C in all cases and it would be of interest to know whether cooling the reaction mixture would lead to further improvements.

From an industrial point of view, large scale processors, capable of dealing with up to 200 min^{-1} min^{-1} have been available for a number of years and are already in use for degassing liquids, dispersion of solids, formation of emulsions and large scale cell-disruption.

Suslick has published a rough comparison between typical sonochemical and photochemical efficiencies which demonstrates that homogeneous sonochemistry is typically more efficient than photochemistry and that heterogeneous sonochemistry is several orders of magnitude better [3] (Table 2).

This is of great importance when considering the potential for application of sonochemistry to industrial processes given the current degree of interest in this rapidly expanding field of chemistry.

A summary of the points raised in Chap. 2 is given in the flowchart on page 27. This presents a concise guide to the factors which should be bourne in mind when deciding on the type of instrumentation necessary to tackle a particular problem.

It should be noted that most of the commercially available systems were not designed with the synthetic chemist in mind. However, most can be readily adapted using the modifications outlined and the current level of interest suggests that it is only a matter of time before specialised instrumentation becomes widely available.

3 Aqueous Sonochemistry

The effects of ultrasound on aqueous solutions of reagents has been exhaustively investigated, mostly to little end, from a synthetic point of view. Much of the early work was concerned with observations that a variety of inorganic substrates underwent not only oxidation, but also reduction reactions when exposed to ultrasound. For example:

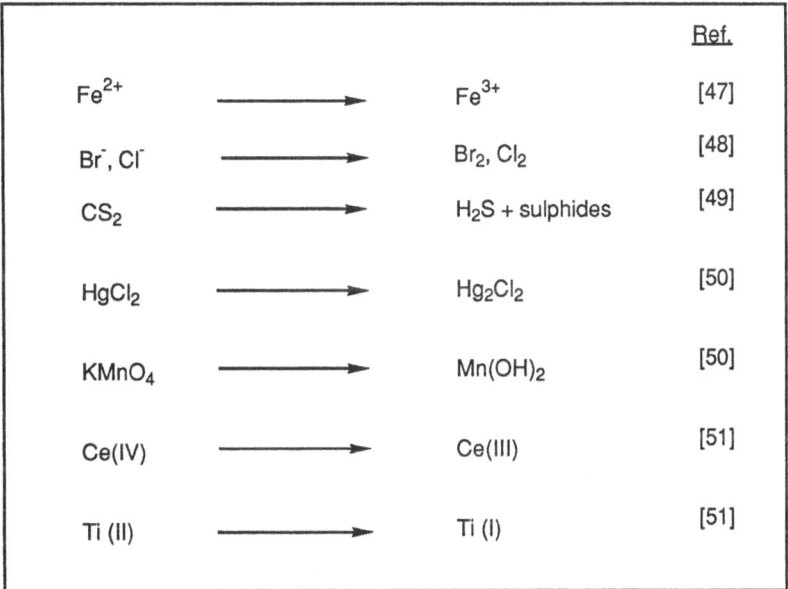

			Ref.
Fe^{2+}	$\longrightarrow$	Fe^{3+}	[47]
Br^-, Cl^-	$\longrightarrow$	Br_2, Cl_2	[48]
CS_2	$\longrightarrow$	H_2S + sulphides	[49]
$HgCl_2$	$\longrightarrow$	Hg_2Cl_2	[50]
$KMnO_4$	$\longrightarrow$	$Mn(OH)_2$	[50]
$Ce(IV)$	$\longrightarrow$	$Ce(III)$	[51]
$Ti(II)$	$\longrightarrow$	$Ti(I)$	[51]

Scheme 4

This arises as a direct result of the ease with which the water molecules fragment in the presence of ultrasound. Careful analysis of the products obtained has shown that a number of highly reactive species are produced: (Scheme 5).

Hence, sonolysis with an immersion horn under typical laboratory conditions will produce hydrogen peroxide at a rate of about 30 µM/minute. This appears to arise from combination of OH and H radicals whose existence have now been conclusively determined by spin trapping experiments [52]. In addition, Margulis has proposed the formation of solvated elec-

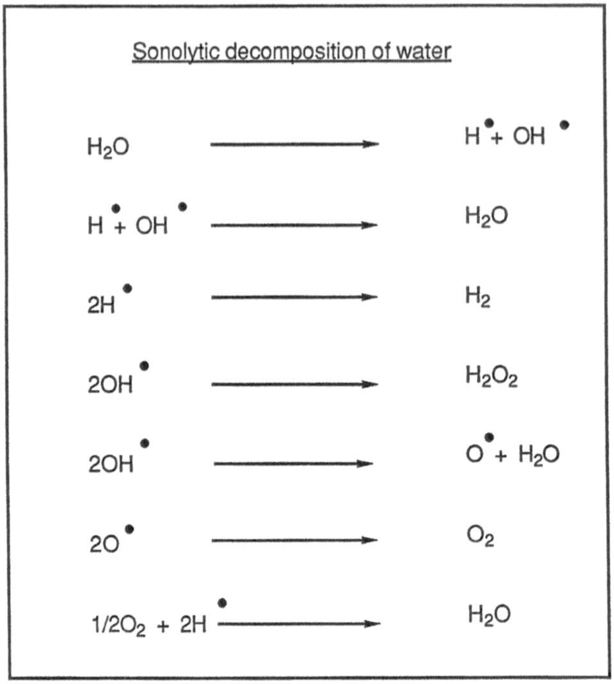

Sonolytic decomposition of water

$$H_2O \longrightarrow H^{\bullet} + OH^{\bullet}$$

$$H^{\bullet} + OH^{\bullet} \longrightarrow H_2O$$

$$2H^{\bullet} \longrightarrow H_2$$

$$2OH^{\bullet} \longrightarrow H_2O_2$$

$$2OH^{\bullet} \longrightarrow O^{\bullet} + H_2O$$

$$2O^{\bullet} \longrightarrow O_2$$

$$1/2O_2 + 2H^{\bullet} \longrightarrow H_2O$$

Scheme 5

trons under conditions of high pH [53], although this has subsequently been refuted by Henglein [29].

Reaction in the presence of such high energy species would be expected to be nonspecific and dominated by secondary chemical reactions unrelated to the direct process, and indeed this is found to be the case. In most cases, sonication of aqueous solutions of organic substrates results in degradation. For example, the only products isolated from sonolysis of a variety of amino acids, and proteins were formaldehyde, primary amines, hydrogen, carbon monoxide and ammonia. Similar results were obtained from a range of carbohydrates. From a synthetic point of view, there are very few examples where discrete organic products have been isolated from the reaction mixture. A selection of these, most of which date from the early 1950s, are summarised in Scheme 6. Furthermore details of these investigations are contained in some of the earlier reviews dealing with the chemical effects of ultrasound [26a, 33, 54].

Consideration of the high vapour pressure of water relative to inorganic or dilute organic reagents would predict that these reactions must *always* be dominated by secondary reactions. However, Luche has recently described the apparently paradoxical formation of organozinc and tin species in aqueous solutions [62]. Optimal conditions involve the use of organic co-solvents, however, the reactions can still be executed in pure water (Scheme 7). These results were later corroborated by Pietrusiewicz and Zablocka, who reported

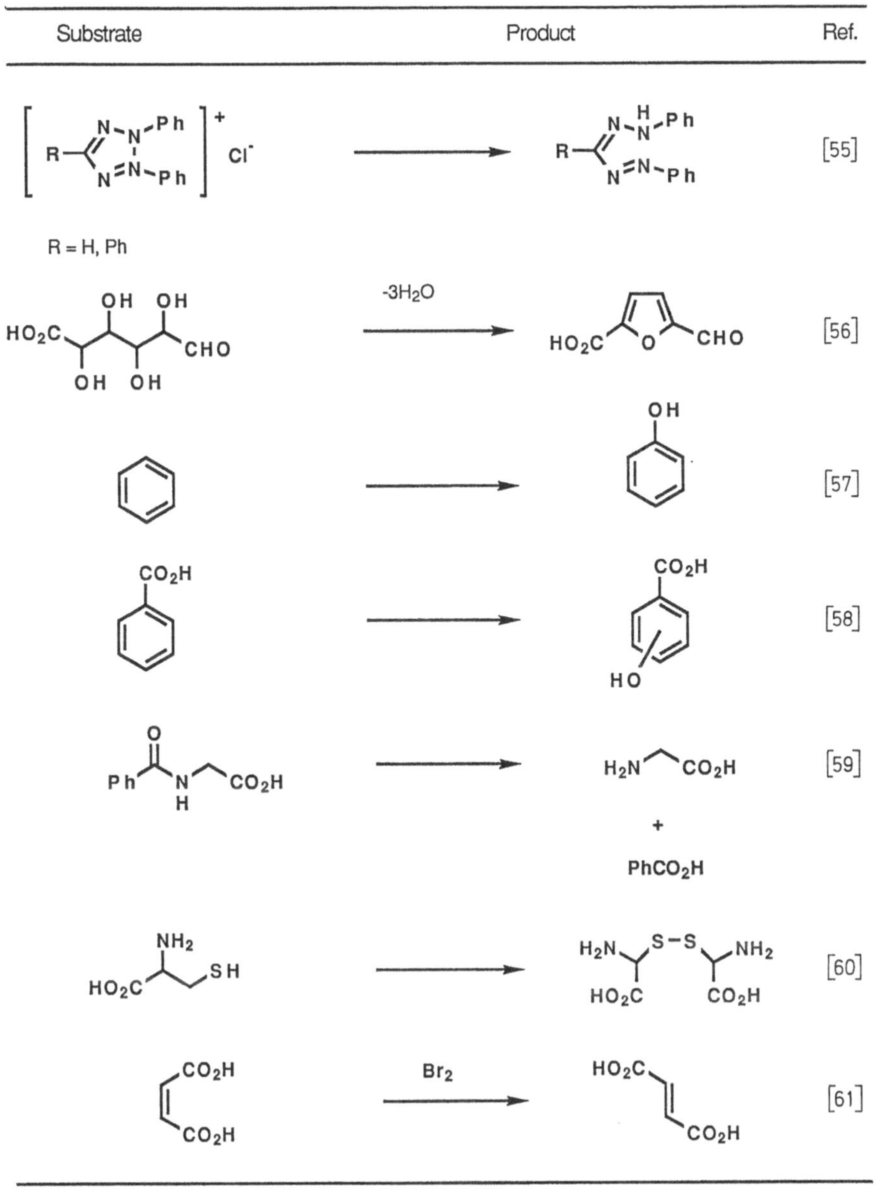

Substrate	Product	Ref.
$\left[R-\overset{N-N-Ph}{\underset{N=N-Ph}{\diagdown}} \right]^+ Cl^-$ R = H, Ph	$R-\overset{N-\overset{H}{N}-Ph}{\underset{N=N-Ph}{\diagdown}}$	[55]
HO_2C OH OH OH OH CHO (−3H₂O)	HO_2C—O—CHO	[56]
(benzene)	(phenol) OH	[57]
CO₂H (benzoic acid)	CO₂H HO	[58]
$Ph\overset{O}{\underset{}{\diagdown}} \overset{}{\underset{H}{N}} CO_2H$	H_2N CO₂H + PhCO₂H	[59]
NH₂ HO₂C SH	H_2N S—S NH₂ HO₂C CO₂H	[60]
CO₂H CO₂H (Br₂)	HO₂C CO₂H	[61]

Scheme 6

that similar species could be used to alkylate vinyl phosphine oxides [63] (Scheme 8).

The nature of the organozinc species has not been elucidated, but it seems likely that the reaction proceeds via a radical pathway, where the presence of water would be beneficial. It should also be stressed that these results are

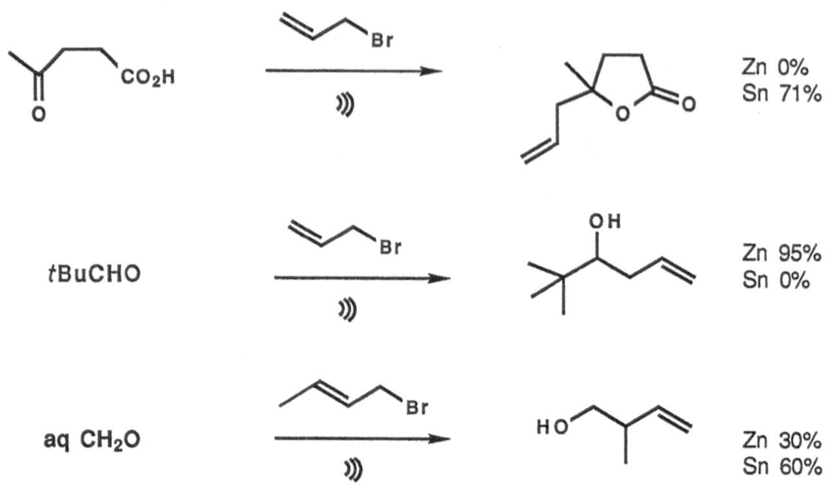

			Zn 0% Sn 71%
tBuCHO			Zn 95% Sn 0%
aq CH$_2$O			Zn 30% Sn 60%

Solvents used: EtOH/H$_2$O (9:1); acetone/H$_2$O (4:1); pyridine/H$_2$O (1:2); pure H$_2$O

Scheme 7

virtually the first of real synthetic use to merge from extensive studies which span the past 30 years.

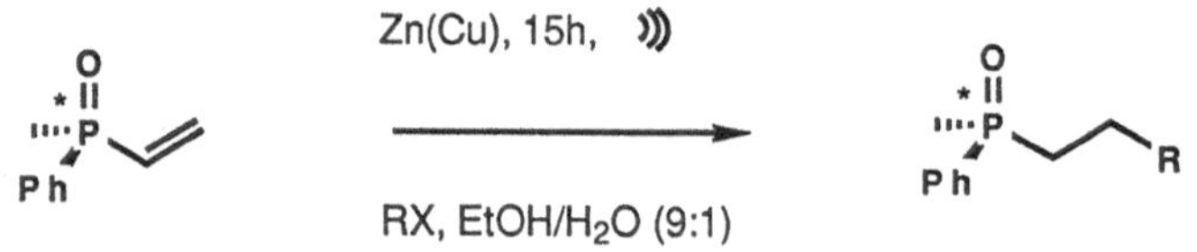

Zn(Cu), 15h,)))

RX, EtOH/H$_2$O (9:1)

Scheme 8

4 Preparation of Activated Magnesium

Grignard reagents are one of the few organometallic species that have been fully integrated into the synthetic chemist's repertoire, and the "tricks" that have been developed to deal with the initiation of the reaction between the organohalide and magnesium are many and varied. These range from the addition of a crystal of iodine, or a few drops of 1,2-dibromoethane to the use of highly coordinating solvents, or freshly-turned magnesium metal. Nevertheless, there are a number of organic halides which continue to give unsatisfactory results under these reaction conditions. As a result, much effort has been invested in examining other methods for activating the metal.

Renaud first reported that improved yields to Grignard reagents, as well as organolithiums and aluminiums, could be obtained by sonolysis of the reagents in undried diethyl ether (960 kHz, 2 W cm^{-2}). These results were first published in 1950 but did not kindle further interest in the use of ultrasound to accelerate chemical reactions [64].

Renaud's initial results were later followed up by Sprich and Lewandos who examined the reaction between magnesium and 2-bromobutane in diethyl ether [65] (Scheme 9).

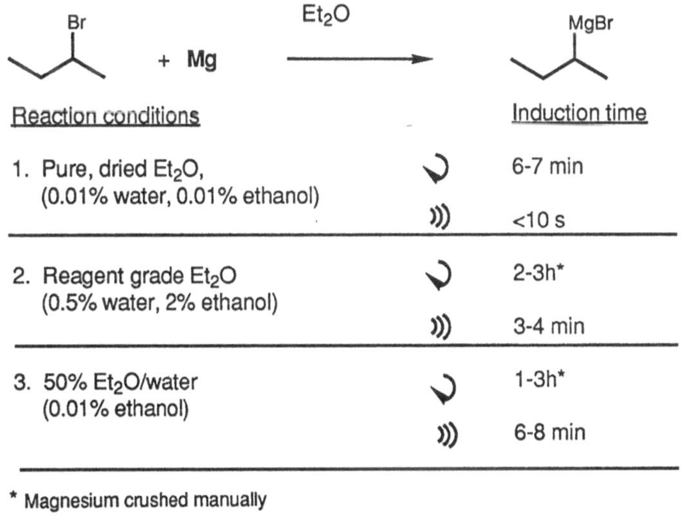

Reaction conditions		Induction time
1. Pure, dried Et$_2$O, (0.01% water, 0.01% ethanol)	↻	6-7 min
	)))	<10 s
2. Reagent grade Et$_2$O (0.5% water, 2% ethanol)	↻	2-3h*
	)))	3-4 min
3. 50% Et$_2$O/water (0.01% ethanol)	↻	1-3h*
	)))	6-8 min

* Magnesium crushed manually

Scheme 9

Ultrasound has a profound effect on the initiation of the reaction although the overall yield of Grignard reagent was unchanged. Sonication was carried out in a cleaning bath and it is significant that sonication of the metal prior to addition of the bromide had no effect on the initiation time. This suggests that the primary effect of the ultrasound is to remove adsorbed water from the metal surface, rather than any surface cleaning effects.

Bonnemann et al. [66] have described a method for the preparation of highly activated magnesium which was used in the reductive synthesis of a variety of transition metal complexes known to be active catalysts for organic reactions. Addition of a small amount of anthracene to a suspension of magnesium powder (particle size <0.1 mm) in THF gives a sparingly soluble adduct, thought to be (anthracene). Mg.3THF. Sonolysis of the pre-formed complex in the presence of the desired reagents at 65 °C releases highly activated magnesium. The anthracene then reacts with more of the magnesium powder, completing the catalytic cycle. That is, its role is effectively one of a phase transfer catalyst (Figure 21).

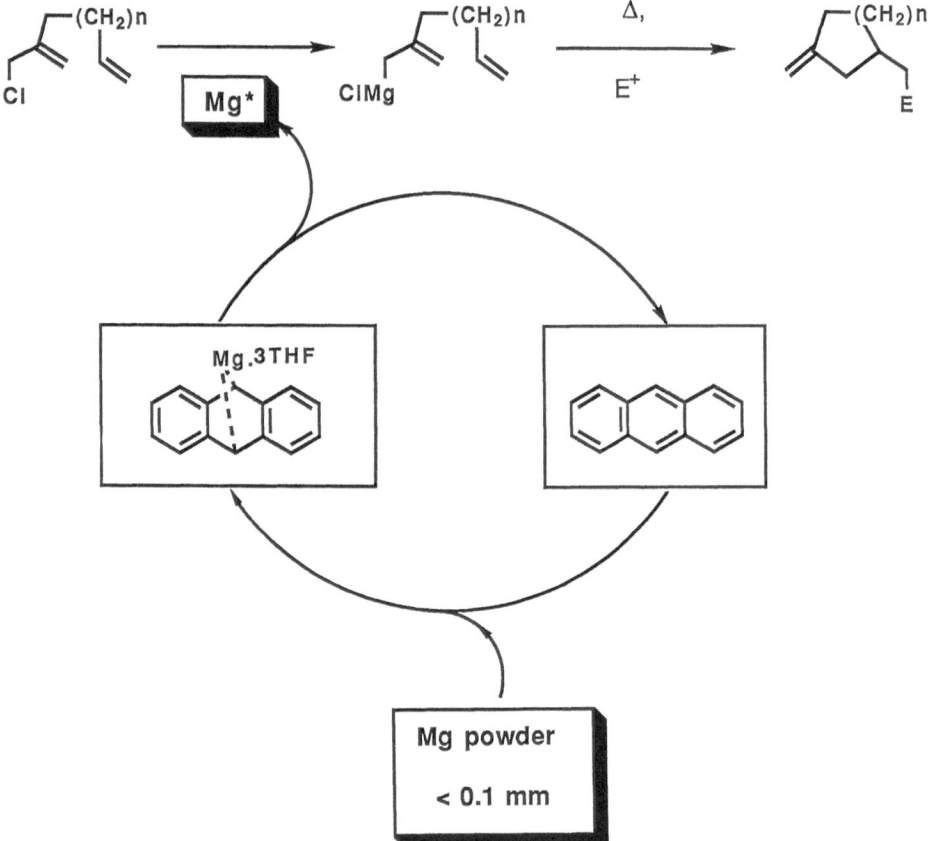

Fig. 21. Preparation and reaction of activated magnesium via the Mg.3THF complex (anthracene).

Oppolzer and Schneider [67] later used this procedure to prepare allyl-magnesium halides when standard procedures proved unrealiable. Of the various methods available for the production of highly activated magnesium this was clearly the method of choice. Its advantages stem from the fact that it obviates the need for the complicated apparatus required to vaporise metallic magnesium [68] or the tedious stoichiometrically precise manipulation employed in the preparation of "Riecke magnesium" [69]. Yields of the cyclised products were high and the undesired reaction of the Grignard reagent with the starting material was suppressed. The methodology was later extended to provide two elegant syntheses of the sesquiterpenes (±)-chokol A [70] (Scheme 10), (±)-6-Protoilludene and its C-3 epimer [71]. The key step in each case was an intramolecular magnesium-ene reaction. It should be noted that the authors reported that reactions were cleaner in the absence of the orange magnesium-anthracene complex, which was removed before addition of the chloride to the activated metal.

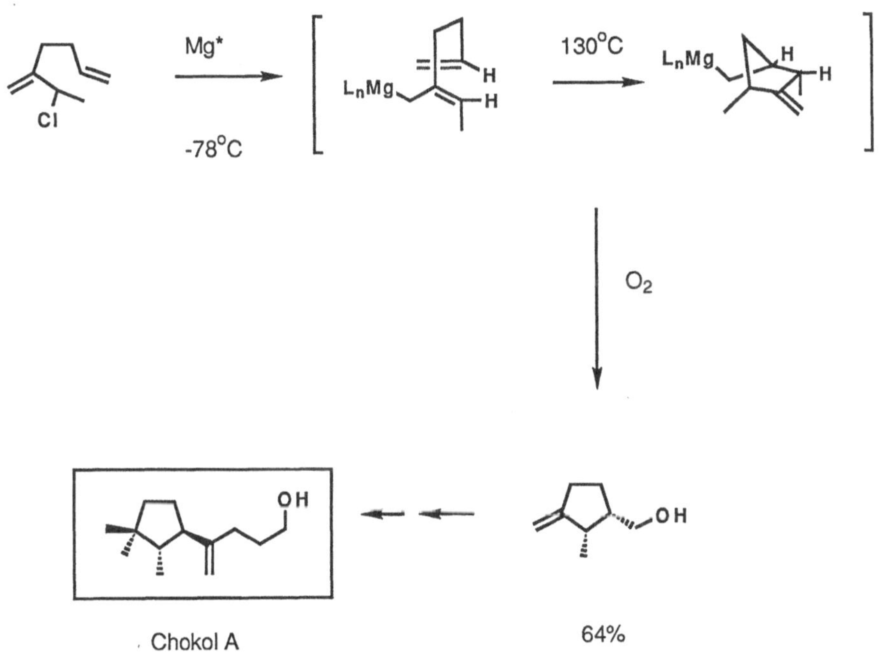

Chokol A   64%

Scheme 10

The uses of the Mg(anthracene) complex have been extensively studied by Bogdanovic and co-workers. For example, it is known to mediate hydrogenation reactions in the presence of a transition metal halide (e.g. $CrCl_3$ [72], $TiCl_4$). The primary process is the formation of the true catalyst, followed by dissociation to give quasi-atomic magnesium. Hydrogenation reactions proceed smoothly at 60 °C and 60 bar [73] (Scheme 11).

35

Mg(anthracene)
+
MXn
[MMgX]
+
anthracene
[MMgX]
Mg
+
H2
MgH2
anthracene
Δ

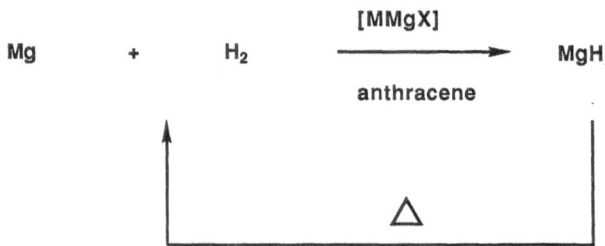

Scheme 11

The activity of the magnesium hydride prepared in this way is contrary to that available by literature methods and the reagent has been used as a reducing agent in its own right (Scheme 12).

$$SiCl_4 \xrightarrow{2MgH_2} SiH_4 + 2MgCl_2$$

Scheme 12

Furthermore, reaction with olefins in the presence of zirconium tetrachloride produces dialkylmagnesiums [75] (Scheme 13).

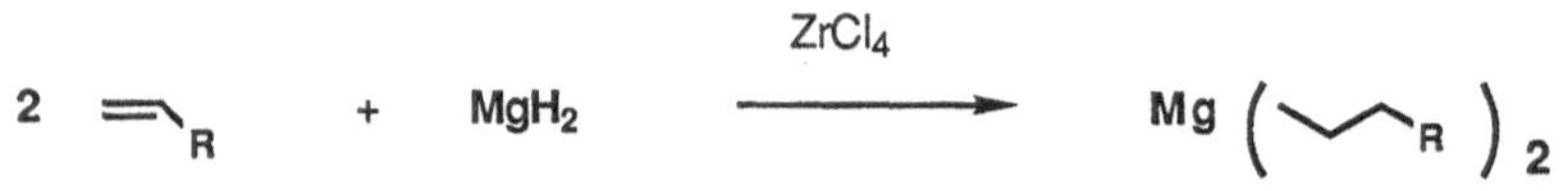

2
R
+
MgH2
ZrCl4
Mg
R
2

Scheme 13

If the hydrogenation reaction is carried out in the presence of magnesium chloride, the product is the saturated Grignard reagent. The active species is HMgCl, which is formed quantitatively [76] (Scheme 14).

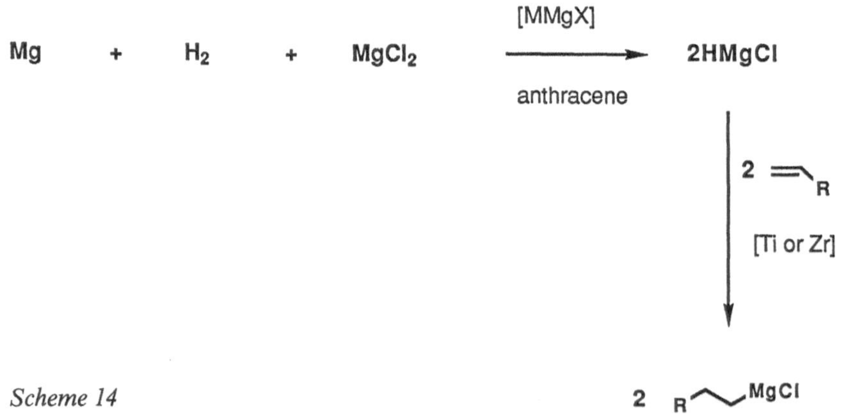

Mg
+
H2
+
MgCl2
[MMgX]
anthracene
2HMgCl
2
R
[Ti or Zr]
2
R
MgCl

Scheme 14

A variety of transition metal organometallics are also available in a one-pot process [76] (Scheme 15). The products obtained are more usually prepared using aluminium alkyls or sodium borohydride.

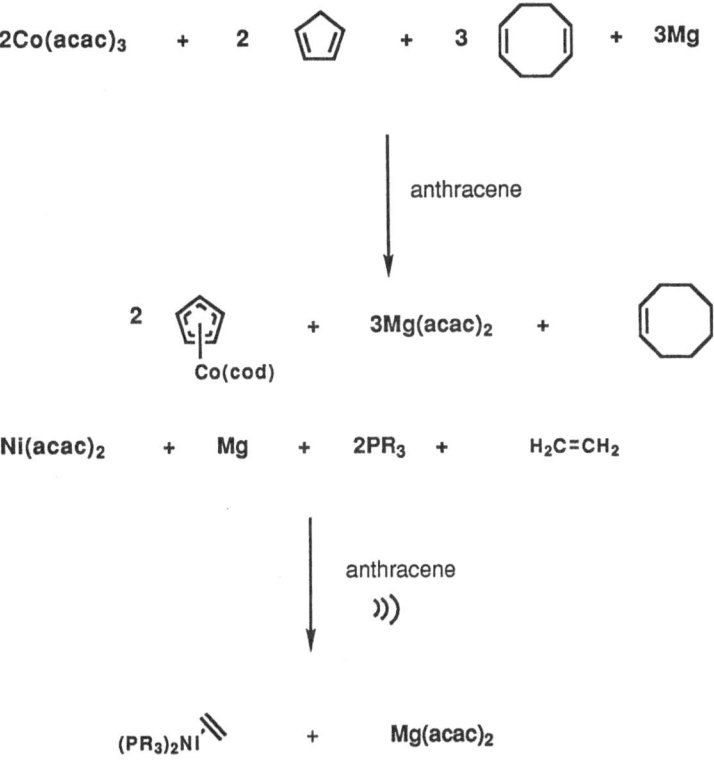

Scheme 15

Highly active, finely divided magnesium powder can also be prepared by the two-step process shown [77] (Scheme 16).

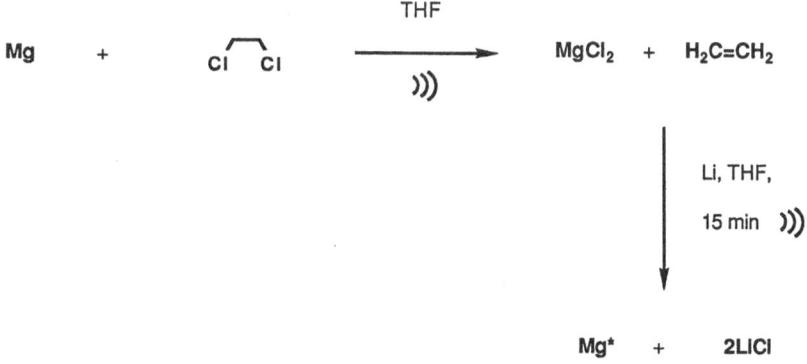

Scheme 16

The overall process is rapid and avoids the need to use the high boiling solvents recommended by Rieke [69a], or naphthalene, which may prove difficult to remove. *p*-Bromoethylbenzene reacts with magnesium prepared in this way at 0 °C and hydrolysis of the Grignard reagent gives the reduced product in over 90% yield. These results are comparable with those of Riecke.

Reduction of α,β-unsaturated amides, lactams, nitriles, and ketones using activated magnesium in methanol has been reported [78]. The reaction proceeds in good yield and is highly specific for the conjugated double bond in most cases. However, the length of the induction period ranges from minutes to several hours. In addition, it is highly exothermic, confining its use to small scale reactions. Interest in this reaction was prompted by the failure of a rhodium catalysed hydrogenation to effect reduction of an eneaminoindole intermediate used in the synthesis of (±)-indolactam V [79], one of a family of compounds which possess potent tumour promoting and skin vesicatory properties.

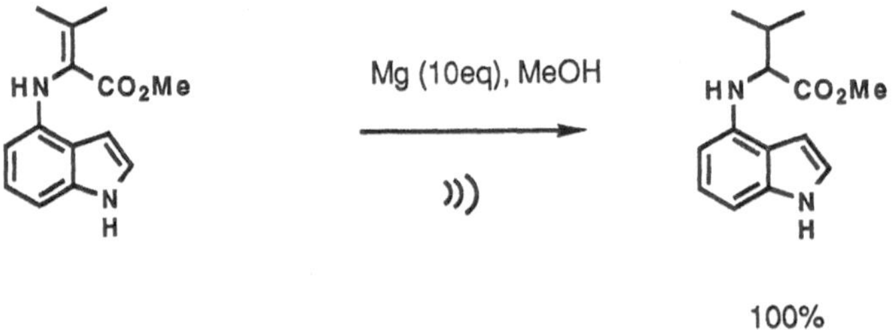

Scheme 17

Remarkably, the reaction proceeded smoothly and gave a quantitative yield of the aminoindole, uncontaminated by products arising from reaction with the indole nucleus (Scheme 17). Moreover, reaction was instantaneous and occurred in a controlled manner — an important consideration in view of the fact that this was the second step of an eleven-stage synthesis.

5 Preparation of Organoaluminium Compounds

The first paper describing the application of ultrasound to the preparation of main group organometallic compounds detailed a preparation of trialkylaluminium compounds by direct reaction of Grignard reagents formed in situ with aluminium powder [64]. Activation of the aluminium, which generally involves use of Al—Mg alloys, proved unnecessary. Recent work has shown that organoaluminium species of the type $R_3Al_2X_3$ are available via reaction of alkyl halides with aluminium powder [80] (Scheme 18), although yields of subsequent reactions are poor.

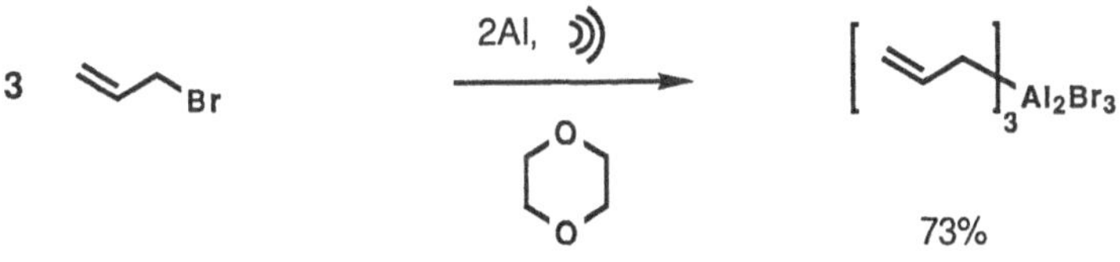

Scheme 18

However, a slight modification of this procedure showed that trimethylaluminium was available in 86% yield and 95% purity [81]. Methylaluminium sesquiiodide was formed by direct reaction of aluminium powder with methyl iodide in ethanol. Subsequent treatment with triethylaluminium gave the triethyl derivative.

Furthermore, reaction of ethylaluminium sesquiiodide with trialkylborates produces triethylborane in high yield [82]. The thermal reaction at 36 °C takes twice as long. Other alkyl borates were examined, but all gave lower yields of triethylborane (Scheme 19).

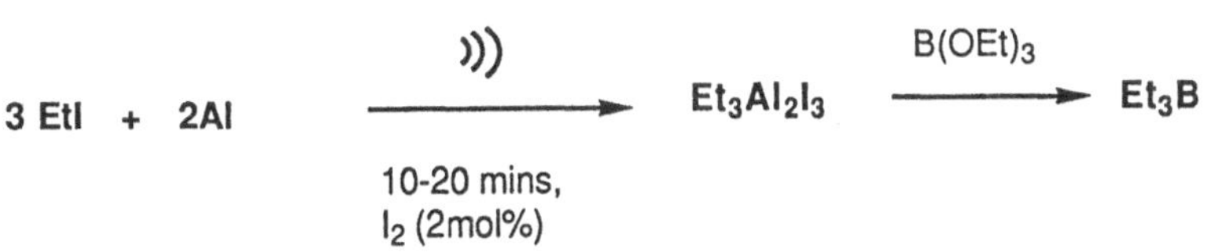

Scheme 19

Brown has shown that ultrasound can be used to accelerate the classic magnesium mediated reaction, allowing access to a number of organoboranes which are not available by hydroboration [76].

6 Application of Ultrasound to the Preparation of Organolithiums

6.1 Preparation of Organolithium Reagents

Luche and Damiano initially reported that sonication of alkyl or aryl halides and lithium gave THF-solutions of organolithium reagents in 1980 [83]. The reaction was shown to be generally applicable and excellent yields of *n*-propyl, *n*-butyl, and phenyllithiums could be obtained (61–95%) even in wet technical grade THF! Reaction with isopropyl and *t*-butyl halides was sluggish, but this reaction should hold potential for large scale industrial applications. Its advantages lie in its rapidity and efficiency, allowing instantaneous initiation of the reaction. This is especially important when working on a large scale where the unpredictability of the initiation period can present a serious problem.

It was later shown that this methodology was equally applicable to the in situ preparation of the commonly used non-nucleophilic base lithium diisopropylamide (LDA) [84] (Scheme 20). The one-pot reaction allows preparation of small (<10 mmols) quantities of LDA/THF within 15 min, although reaction on a 0.5 M scale took proportionally longer (2 h).

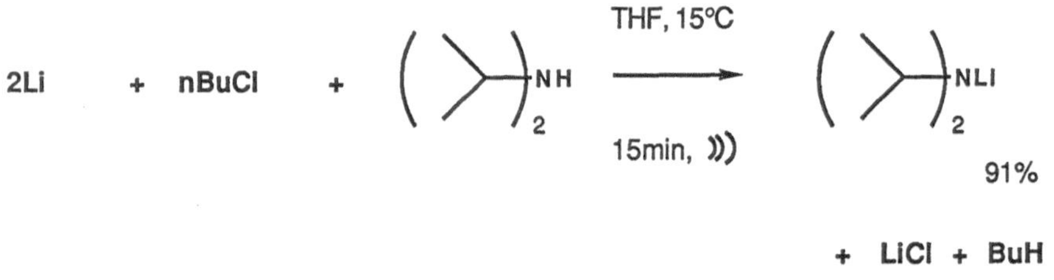

Scheme 20

Reaction of diisopropylamine with the butyllithium formed in this way appears to be virtually instanteous as no side-products arising from attack of *n*BuLi on the carboxyl function of isobutyric acid were observed in the one-pot reaction shown below:

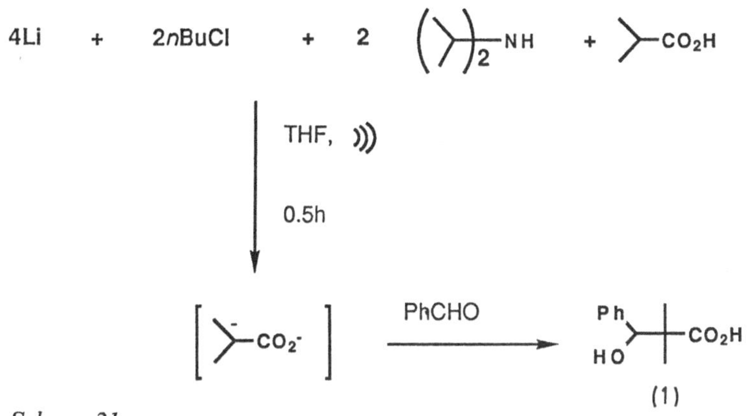

4Li + 2nBuCl + 2 $\left(\!\!\!\!\begin{array}{c} \\ \end{array}\!\!\!\!\right)_2$-NH + $\rangle$-CO₂H

THF,)))

0.5h

$\left[\rangle\!\!-\text{CO}_2^-\right]$ →PhCHO→ Ph, $\!\!\begin{array}{c}\text{HO}\end{array}$-CO₂H

(1)

Scheme 21

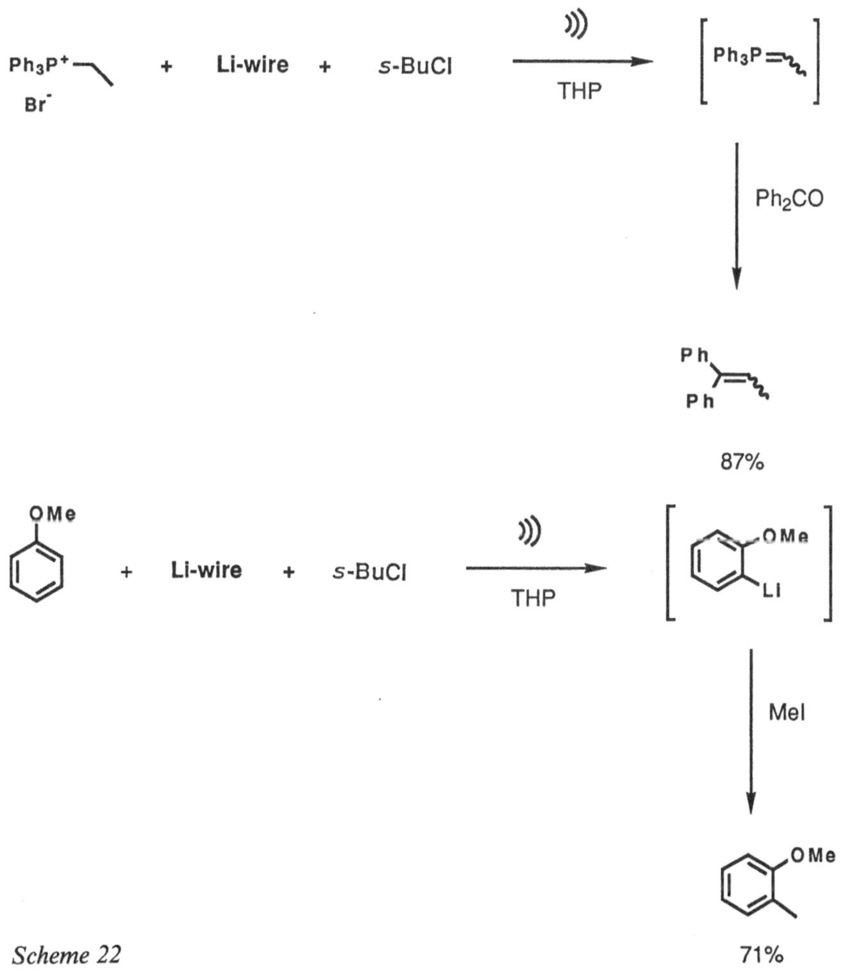

Ph₃P⁺—\\ + Li-wire + s-BuCl →)))—THP→ $\left[\text{Ph}_3\text{P}=\!\!\!\backsim\right]$

Br⁻

Ph₂CO

Ph\\
Ph—

87%

OMe

+ Li-wire + s-BuCl →)))—THP→ $\left[\begin{array}{c}\text{OMe}\\\text{LI}\end{array}\right]$

MeI

OMe

Scheme 22   71%

Quenching the dianion with benzaldehyde gave the β-hydroxyacid (*1*) as the sole product (Scheme 21).

The reactivity of the ultrasonically formed butyllithiums was also examined. Thus, sec-BuLi was used to deprotonate a phosphonium salt and the resulting ylide used in the standard Wittig reaction. Similarly, *ortho*-lithiation of anisole could be effected in a one-pot process (Scheme 22). However, deprotonation of terminal alkynes or 1,3-dithianes requires use of lithium containing at least 2% sodium, in contrast to the previous examples, which were all carried out using low-sodium content lithium wire (<0.02%). There is no satisfactory explanation for this observation as yet.

In situ formation of organolithiums using the methodology outlined has been put to good use in the generation of chloromethyllithium [85]. Reaction with a variety of aldehydes and ketones gave the relevant epoxide by intramolecular cyclization of the intermediate chloride (2) (Scheme 23).

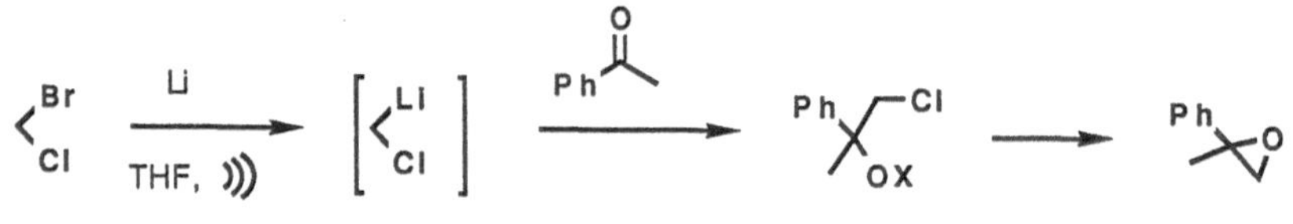

Scheme 23   (2) X = Li, or H

Initial attempts using a cleaning bath as the source of ultrasound gave lower yields of epoxide than the equivalent stirred reaction. This was probably due to further reaction of the epoxide under the conditions used. This problem was circumvented by using a probe system to control the energy input whilst reducing the reaction temperature. Hence, the authors reason that the rate of formation of chloromethyllithium was increased with respect to the rate of cyclisation. That is, formation of the epoxide becomes the rate-determining step. Optimal conditions involved sonication of the reaction mixture for 20 minutes at −15 °C in the case of ketones and −50 °C for aldehydes (Table 3). Cyclization to give the epoxide was generally spontaneous, although the chlorohydrin could be isolated in a number of cases (Scheme 24).

The simplicity of the one-pot procedure compares favourably with exist-

Table 3

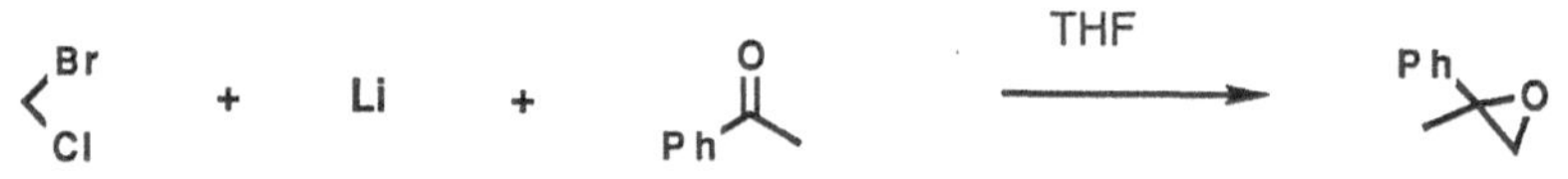

Reaction conditions	T/°C	% GC Yield
stir, 4 h	20	55
ultrasound bath, 0.7 h	15	0–35
ultrasound probe, 0.33 h	20	72
	−15	98

ing methods for the generation of epoxides from carbonyl compounds. Reaction with dibromomethane gave slightly lower yields of epoxide, although it was impossible to obtain successful results using dichloromethane.

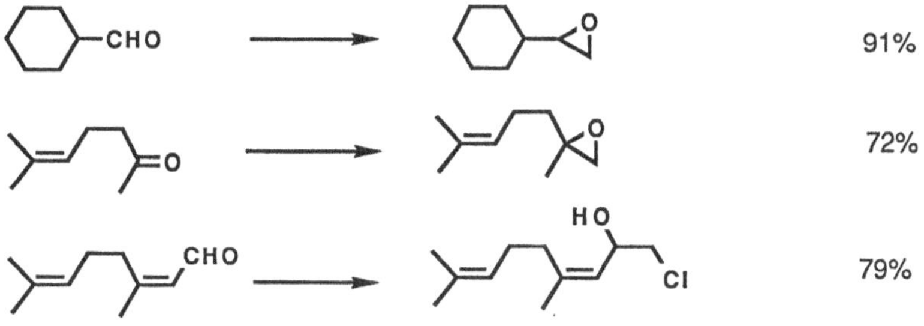

91%

72%

79%

Scheme 24

6.2 Reactions of Organolithiums with Carbonyl Compounds

Reaction of the ultrasonically generated alkyllithiums with carbonyl compounds provides an extremely facile route to secondary and tertiary alcohols in a lithium-mediated Barbier-type reaction.

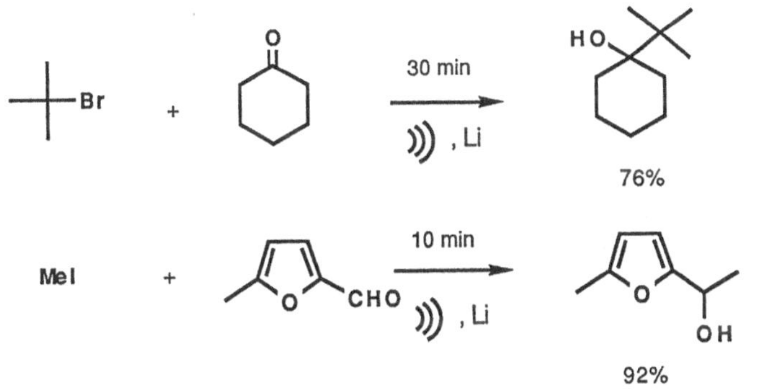

76%

92%

Furthermore, the commonly observed side reactions, such as reduction and enolization, were minimal. The reaction also proceeded cleanly with allylic, vinylic and even benzylic halides, where Wurtz coupling normally predominates.

A number of methyl-substituted 1,2-diphenylethanols have been prepared in this way for use as convenient precursors of the relevant stilbenes [86] (Scheme 25).

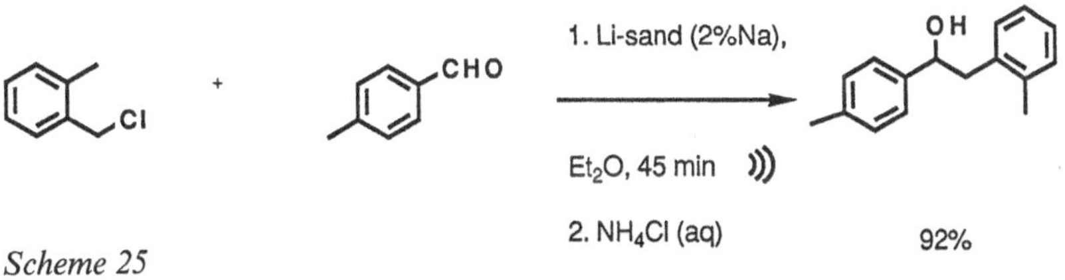

1. Li-sand (2%Na),

Et₂O, 45 min))

2. NH₄Cl (aq)

92%

Scheme 25

Optimising the reaction conditions showed that diethyl ether was a better solvent than THF, as recommended by Luche in his original publication. Furthermore, the presence of 2% sodium in the lithium sand gave better results than those obtained using commercial lithium wire of varying purity.

Trost and Coppola [87] have put this methodology to good use in an intramolecular synthesis of *exo*-methylene cyclopentanes (**3**) (Scheme 26).

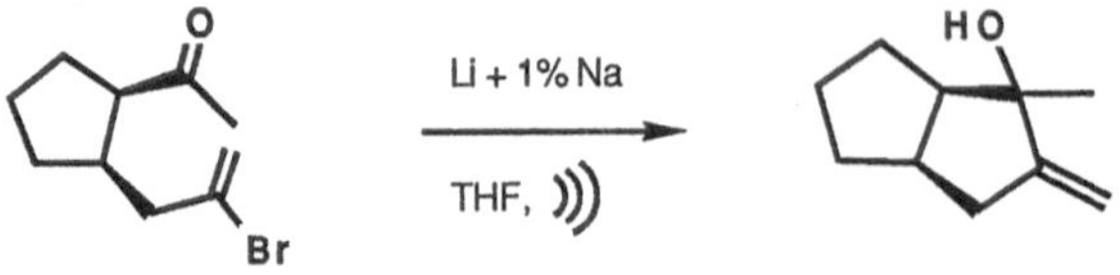

Scheme 26   (3) 95%

In the examples given, the cyclization proceeded with retention of configuration at the centres indicated. This includes the synthesis of the highly strained *trans*-fused system, although competing dehydrobromination reduced the yield to 44%.

Similarly, a lithium-mediated reaction gives the tertiary alcohol (**4**). This constitutes the first step in the total synthesis of (±)-methylpentalenic acid via an intramolecular double Michael reaction [88] (Scheme 27).

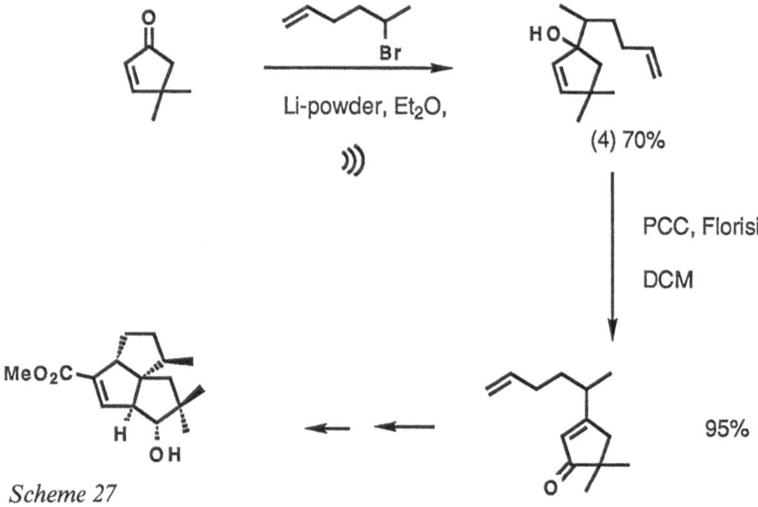

Scheme 27

In an extension of their original work, Luche and co-workers [89] attempted to generate organocopper species in a one-pot reaction. However, 1,2-addition of the intermediate organolithium proved faster than the reaction with the copper iodide. Nevertheless, modifying the reagents used and increasing the power of the ultrasound generator allowed formation of the desired organocopper species at −40 °C within 10 to 30 mins. Addition of the enone and sonication for a further 10 min gave the 1,4-alkylated product (**5**). The reaction is high yielding and extremely fast although it suffers from the need to use C_5H_7Cu-2HMPA [90] as the source of copper.

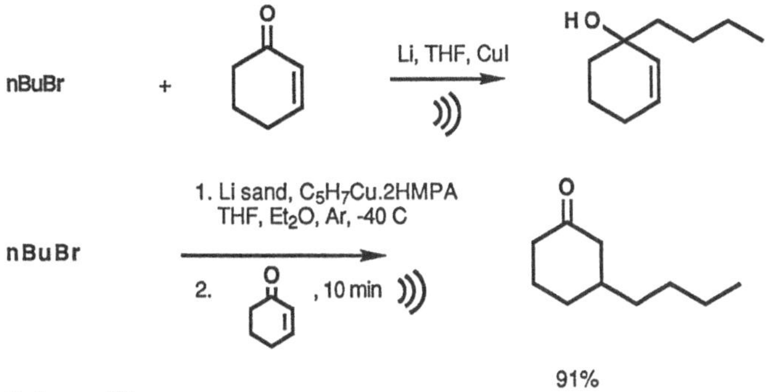

91%

Scheme 28

Araki and Butsugan have recently described the lithium-mediated allyla-tion of carbonyls by means of an allylic phosphate anion [91].

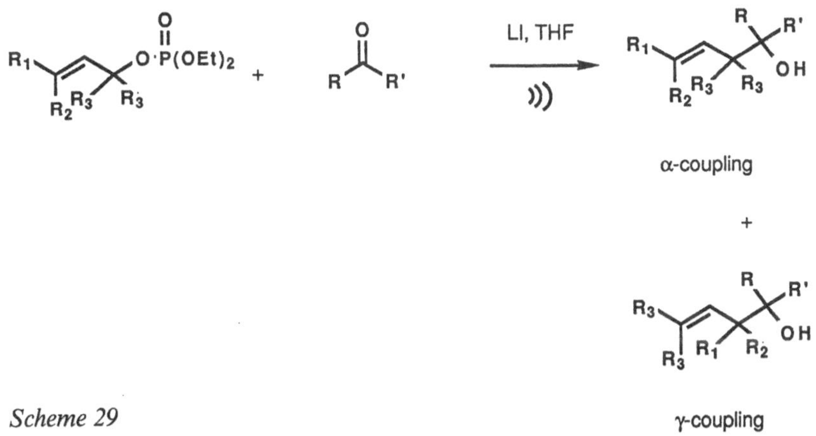

α-coupling

+

Scheme 29   γ-coupling

In general, reactions were carried out using a 2-fold excess of phosphate and good yields of coupled product were obtained using a wide range of carbonyl compounds that included aldehydes, ketones, esters and acid chlor-ides. Furthermore, addition to α,β-unsaturated carbonyl compounds gave the 1,2-addition products specifically. However, the regioselectivity of the process was poor and mixtures of α- and γ-coupled products were isolated in each case (Scheme 29).

The introduction of allyl groups has also been achieved via a related orga-notin reaction [92]. Allyl stannanes were prepared by the Mg-mediated cross coupling reaction (Scheme 30). In most cases, the product of the reaction was isolated in virtually quantitative yield and subsequent reactions could be carried out without the need for further purification. In contrast, Wurtz-type coupling products predominate if the reaction is carried out in the absence of the allyl chloride [93].

These results are corroborated by Boudjouk and Han's study of the equivalent lithium-mediated reaction [94, 95] with both trialkyltin and silicon coupounds. Nevertheless, the cross-coupling reaction occurs faster than Wurtz-type coupling in this case, and products were uncontaminated by

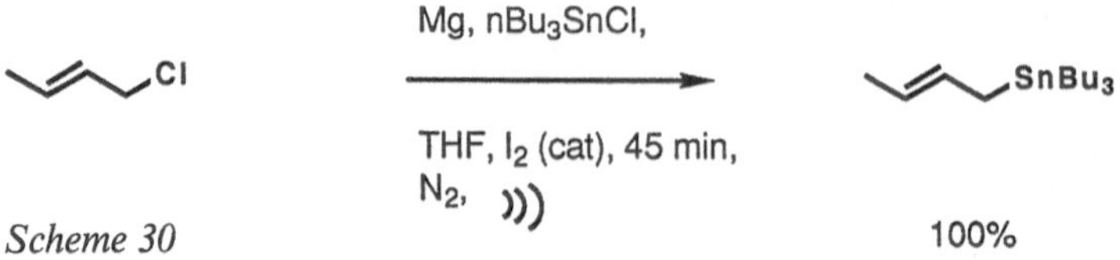

Mg, nBu$_3$SnCl,

$$\text{\cancel{}}\diagup\diagdown\text{CI} \longrightarrow \text{\cancel{}}\diagup\diagdown\text{SnBu}_3$$

THF, I$_2$ (cat), 45 min,

N$_2$,)))

Scheme 30   100%

dialkylstannanes. Furthermore, the reaction occurs with retention of the original stereochemistry in the first instance, although subsequent thermal rearrangement of the product was noted in a number of cases. The ease with which allylstannanes can be obtained by this process contrasts with previously published procedures in which yields are commonly reduced due to the instability of the product to the acidic work-up conditions employed. This should extend the synthetic utility of allylstannanes and the scale on which such reactions can be carried out.

Turning their attention to the Bouveault reaction, Luche et al. found that aldehydes could be generated from alkyl halides in a one-pot reaction (Scheme 31).

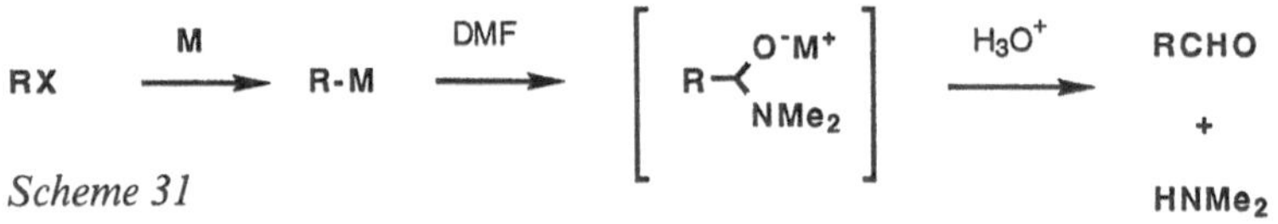

$$RX \xrightarrow{\quad M \quad} R\text{-}M \xrightarrow{\quad DMF \quad} \left[R\diagdown\diagup\begin{array}{l} O^-M^+ \\ NMe_2 \end{array} \right] \xrightarrow{\quad H_3O^+ \quad} \begin{array}{l} RCHO \\ + \end{array}$$

Scheme 31   HNMe$_2$

The classic reaction suffers from numerous side reactions and formation of the aldehyde does not always predominate. However, sonolysis of the reaction mixture gives good yields of uncontaminated aldehyde within 5–10 min using a wide range of 1°, 2° and 3° aliphatic and aromatic chlorides and bromides. In contrast, the thermal reaction at −15 °C takes

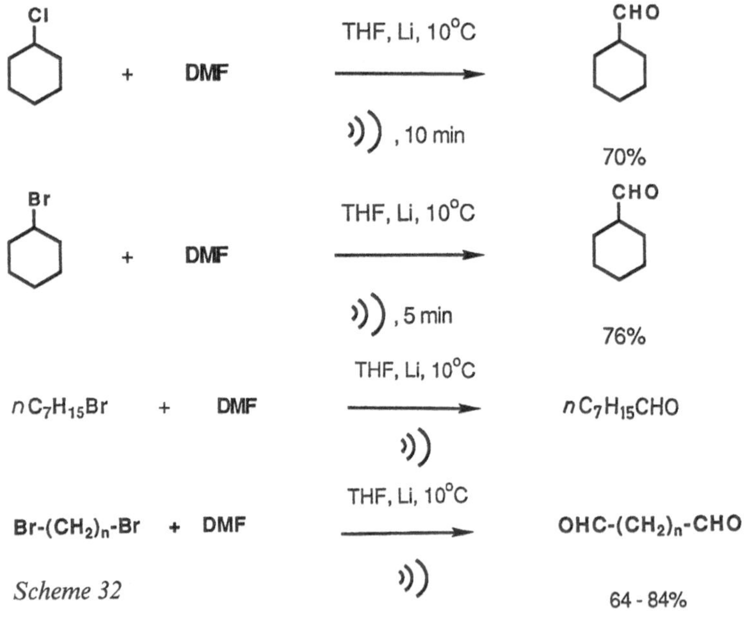

THF, Li, 10°C,))), 10 min → 70%

THF, Li, 10°C,))), 5 min → 76%

nC$_7$H$_{15}$Br + DMF, THF, Li, 10°C,))) → nC$_7$H$_{15}$CHO

Br-(CH$_2$)$_n$-Br + DMF, THF, Li, 10°C,))) → OHC-(CH$_2$)$_n$-CHO

Scheme 32   64 - 84%

considerably longer and yields of aldehyde are lower, particularly in the case of aliphatic substrates [96] (Scheme 32).

Luche et al. have since extended the scope of these reactions by using formamides to direct *ortho*-lithiation of substituted benzenes, thus generating the *ortho*-dialdehyde (**6**) from monohalogenated benzene via a double Bouveault reaction [97, 98] (Scheme 33).

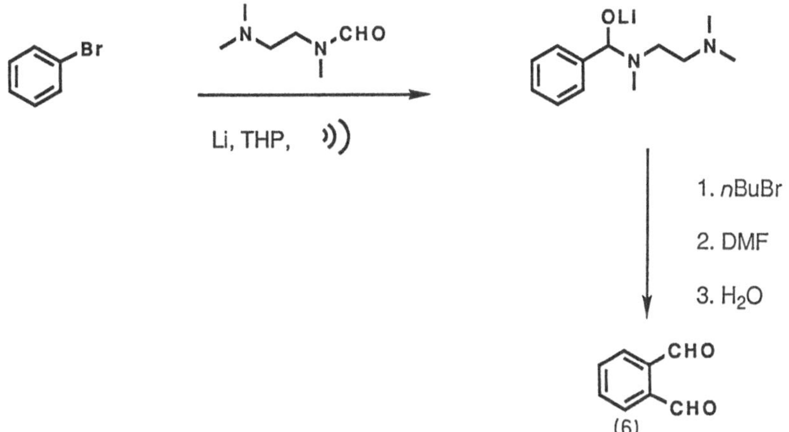

Scheme 33

Comparative experiments with DMF and *N*-formyl-*N'*-methylpiperazine showed that the stability of the α-aminoalkoxide was independent of the nature of the formamide [97], despite reports to the contrary [99]. However, the presence of a second heteroatom directs subsequent *ortho*-lithiation of the aryl ring (Scheme 34).

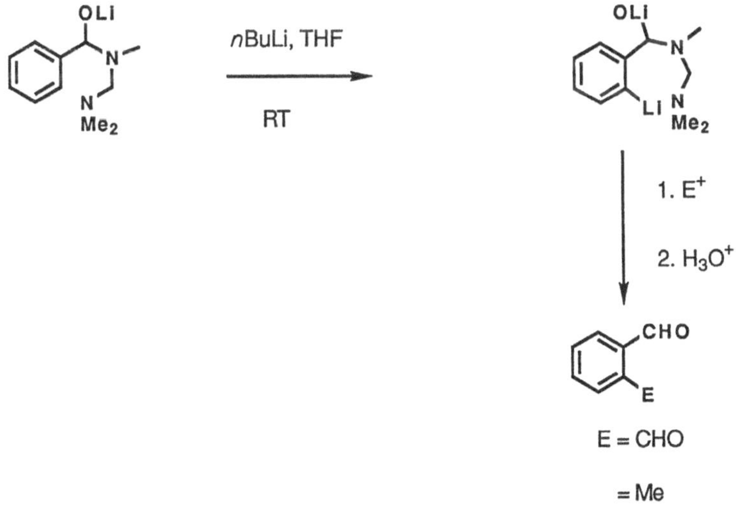

Scheme 34

47

Lithiation proceeds best in tetrahydrofuran solution and quenching with DMF or methyl iodide gave 62% yields of the aldehyde (7). In fact, addition of butyl bromide to the reaction mixture in place of the butyl-lithium gives identical results as the alkylbromide reacted with the excess lithium present to generate the alkyllithium in situ.

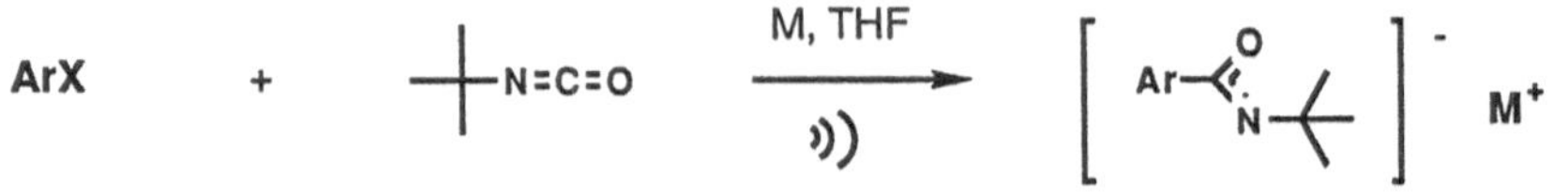

Scheme 35

The α-aminoalkoxides are also available from reaction of aryl halides with alkylisocyanates in the presence of a metal [100] (Scheme 35). However, undirected *ortho*-lithiation of the ring proved much more difficult. Yields of α-aminoalkoxide were highest when magnesium was used but *ortho*-lithiation of the aryl ring required two equivalents of butyllithium as one was consumed in transmetallation. Forming the sodium alkoxide overcame this problem but necessitated the use of a full equivalent of HMPA. If *o*-bromotoluene is treated in this way lithiation occurs at the benzylic position and in some cases the final product can be persuaded to cyclize spontaneously to give the benzo-fused heterocycle (8) (Scheme 36). This methodology has yet to be applied elsewhere; however, it appears to have given a new lease of life to a very old reaction!

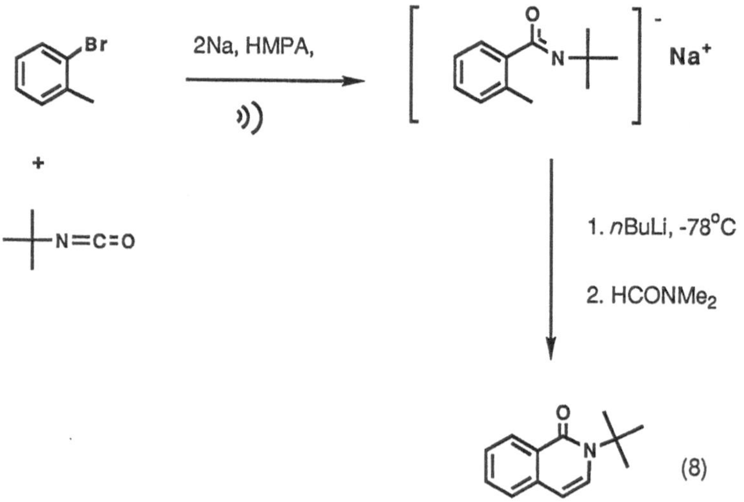

Scheme 36

6.3 Other Reactions of Organolithiums

Other workers have reported that prolonged sonolysis of organolithiums derived from organic halides [94], chlorosilanes and chlorostannanes [95] results in Wurtz-type coupling. Yields are moderate and the reactions are of little synthetic interest. However, coupling of dichlorosilanes and stannanes produces a novel route to the cyclic polysilanes (9) and (10) (Scheme 37). The product obtained is determined by the steric bulk of the alkyl groups and only low levels of contamination by other silanes is observed [95]. Silylene intermediates did not appear to be involved. However, Boudjouk et al. [101] later reported formation of the tetramesityl silylene (11) which had previously been made by photolysis of $(Mes)_2Si(TMS)_2$ [102] (Scheme 38).

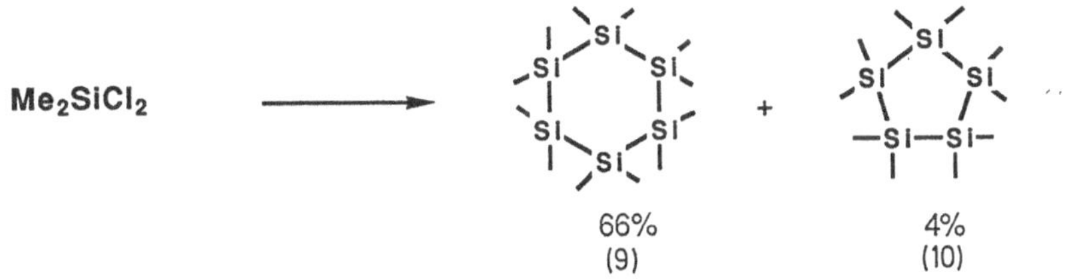

Scheme 37

Sonolysis of the dichloride with lithium in THF was reported to give a 90% yield of the novel silylene. Unfortunately, these results proved irreproducible and Masamune later pointed out that the product generally obtained is the cyclic structure (12) [103].

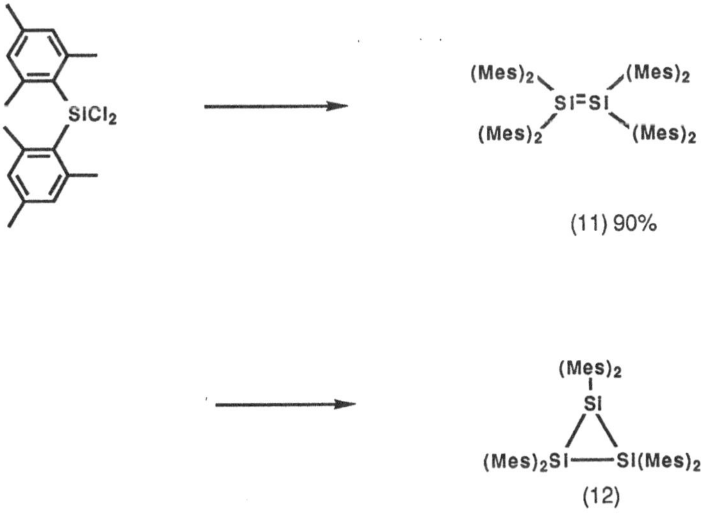

Scheme 38

Lastly, some Chinese workers have developed a simple synthesis of tertiary phosphines based on the reductive cleavage of phosphorous-carbon bonds in phenylphosphines [104, 105] (Scheme 39).

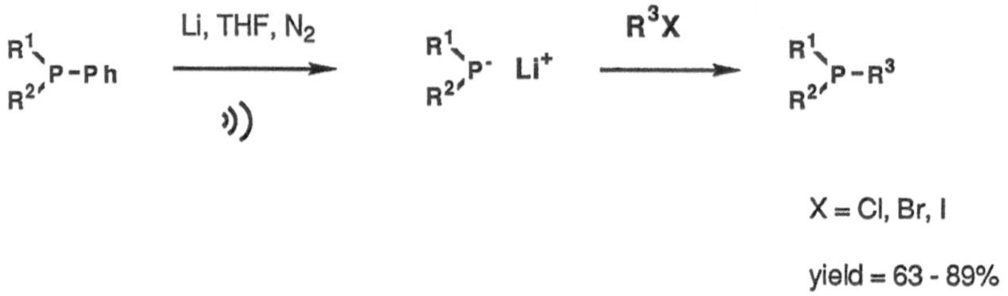

$$R^1_{R^{2'}}\!\!P-Ph \xrightarrow[\text{)))}]{\text{Li, THF, N}_2} R^1_{R^{2'}}\!\!P^- \; Li^+ \xrightarrow{R^3X} R^1_{R^{2'}}\!\!P-R^3$$

X = Cl, Br, I

yield = 63 - 89%

Scheme 39

Reaction was complete within 5 min. In comparison, the stirred reaction takes ten times as long [104]. Alkylation of diphenylphosphines could also be readily achieved to give high yields of uncontaminated product. Previous routes to these important chelating agents had suffered from the extreme sensitivity of the secondary phosphines to air and the difficulty of its exclusion over the long periods of time necessary to obtain satisfactory yields of product.

7 Reactions with Other Alkali Metals

Reactions involving other alkali metals are not as numerous. The properties of colloidal alkali metals have been known for many years but they remain unexploited in synthesis due to the difficulties associated with their preparation. Luche and co-workers observed that small lumps of potassium could be dispersed in a few minutes by sonication in toluene or xylene at 10 °C in a cleaning bath [83]. The colloid generated was used in a number of reactions; for instance, a Dieckman cyclization could be effected within 5 min (Scheme 40).

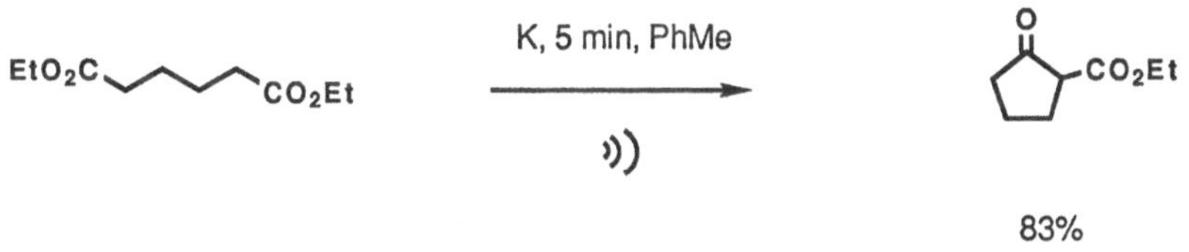

Scheme 40

Interestingly, solvent effects appear to be extremely marked. Luche and co-workers noted that dispersion could not be effected in THF and other reports suggest that the process is extremely sluggish in benzene [106]. Similarly, sodium could only be dispersed in xylene and lithium could not be persuaded to disperse in either of the three solvents tried. Ley et al. have also observed this whilst attempting to form sodium phenylselenide by reaction of sodium with diphenyl diselenide [107]. Using solid sodium, the reaction time was halved when using xylene in place of THF (Table 4).

Luche has suggested that the ease with which metals can be dispersed is related to the lattice energy of the metal. This would explain why

Table 4. Dispersion of alkali metals by sonication in a cleaning bath (15 min, 15–18 °C)

Solvent	Potassium	Sodium	Lithium
Xylene	V	V	X
Toluene	V	X	X
THF			X

lithium, although a ductile metal, will not disperse in any of the above solvents. In fact Luche reports that dispersion cannot even be effected in mesitylene — a highly involatile solvent that would be expected to give highly energetic cavitation [23]. The reduction of (+)-camphor using dispersions of Na, Li and K in THF has also been reported (Scheme 41). The stereochemical outcome of these reactions was identical to those obtained by Birch reduction in liquid ammonia [108a]. These results were used to support their proposal that reduction of enolisable saturated ketones by dissolving metals proceeds via hydrogen transfer within a ketyl dimer in the absence of proton donors [109] rather than that of the previously accepted

Metal	Solvent	Relative % Yield	
K	THF)))	42	58
	NH$_3$	40	60
Na	THF)))	68	32
	NH$_3$	63	37
Li	THF)))	73	27
	NH$_3$	78	22

Scheme 41

dianion mechanism. Interestingly, the authors isolated significant quantities of pinacol when using Na or Li dispersions although this did not present a problem when using K. This is in agreement with Luche's observations

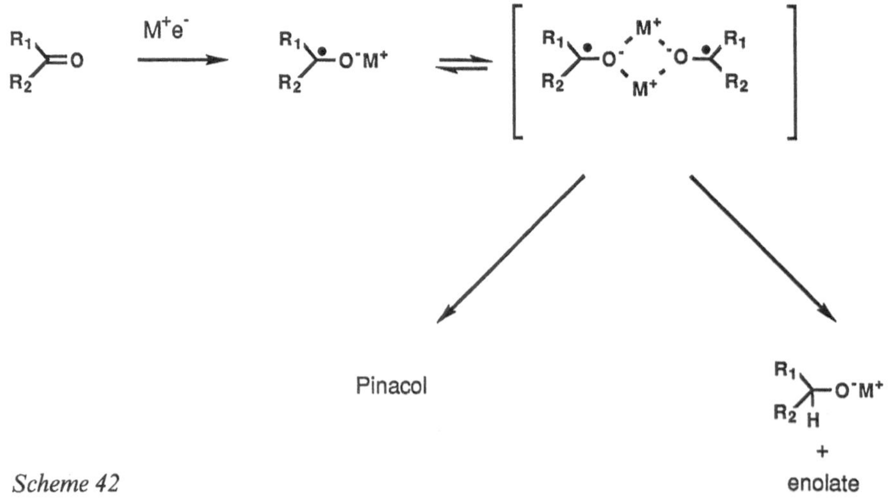

Pinacol

enolate

Scheme 42

that pinacol was a side product of the lithium-mediated reaction between benzaldehyde and heptyl bromide [23]. In addition, sonication of phenols in the presence of lithium and TMS chloride provides an experimentally simple procedure for the synthesis of polysilylated phenols via a modified Birch reduction [108 b].

Colloidal potassium was also used to desulphonylate cyclic sulphones [106]. This reaction is only synthetically useful under these conditions and is extremely slow in the absence of ultrasound. Furthermore, the reaction is highly regioselective and reductive cleavage occurs preferentially at the more substituted centre giving the product shown (13). This contrasts with the low degree of regioselectivity of literature methods (Scheme 43).

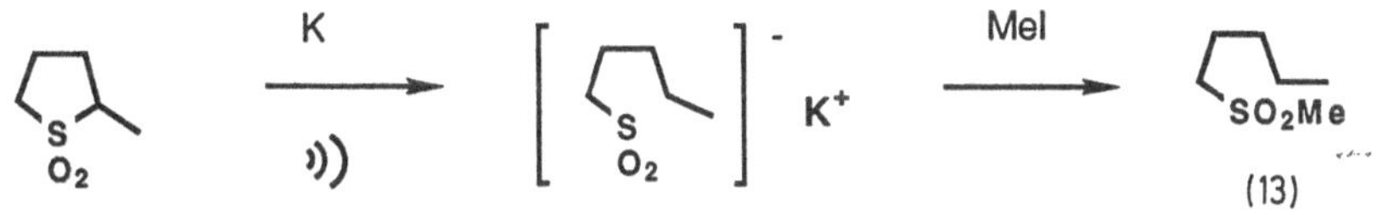

Scheme 43

Similarly, substituted 2-sulpholenes can be regiospecifically ring opened to give the *trans*-γ,δ-unsaturated sulphones. However, the intermediate anion cannot be isolated and must be quenched with methyl iodide to prevent over reduction (Scheme 44).

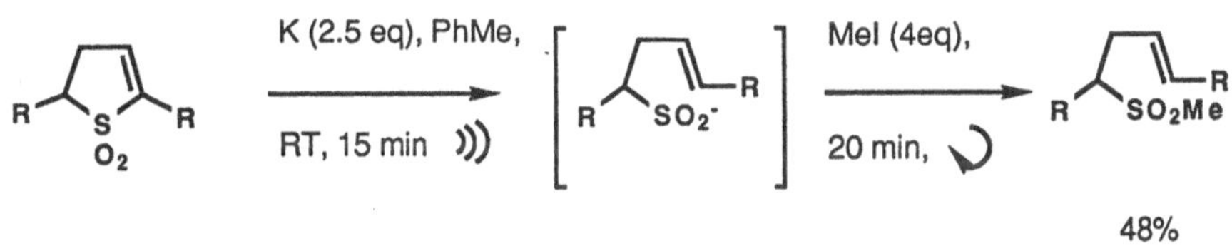

Scheme 44

The authors have proposed that ring opening occurs via the process shown in Scheme 45. Steric considerations would appear to favour ring-opening via intermediate (14a) leading stereospecifically to the *trans* olefin.

3-Sulpholenes are well known precursors of 1,3-conjugated dienes. Chou and You have also shown that ultrasonically dispersed potassium (UDP) can also be used to promote extrusion of sulphur dioxide from the sulpholene, avoiding the need to subject the substrate to high temperature thermolysis. Hence, trans-2,5-dialkyl-3-sulpholenes gave an 8:1 ratio of the (E, Z) and (E, E) dienes in high yield. However, reaction with more sensitive substrates was not attempted [110] (Scheme 46).

Treatment of 4-bromo-2-sulpholenes with UDP results in deprotonation at C5. The product isolated is the bicyclic sulphone (16) [111] which pre-

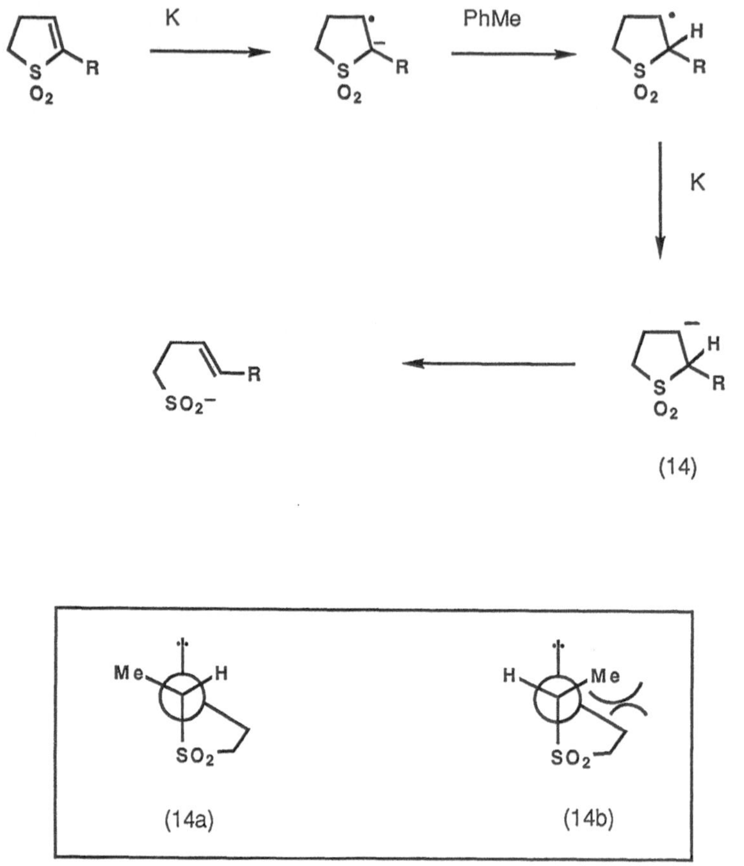

(14)

(14a)   (14b)

Scheme 45

sumably arises from elimination of HBr followed by dimerisation and loss of sulphur dioxide [112] (Scheme 47).

Primary cleavage of the C—Br bond followed by intermolecular abstraction of HBr was ruled out on the basis that the overall conversion of (15)

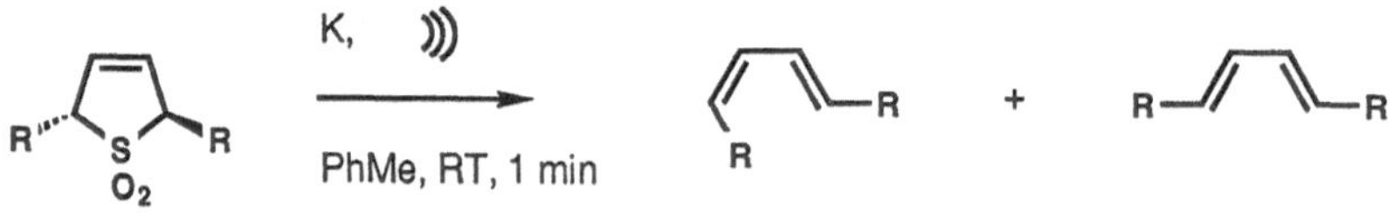

R	% Yield	
n-Pentyl	84	10
n-Hexyl	86	11
n-Heptyl	86	11

Scheme 46

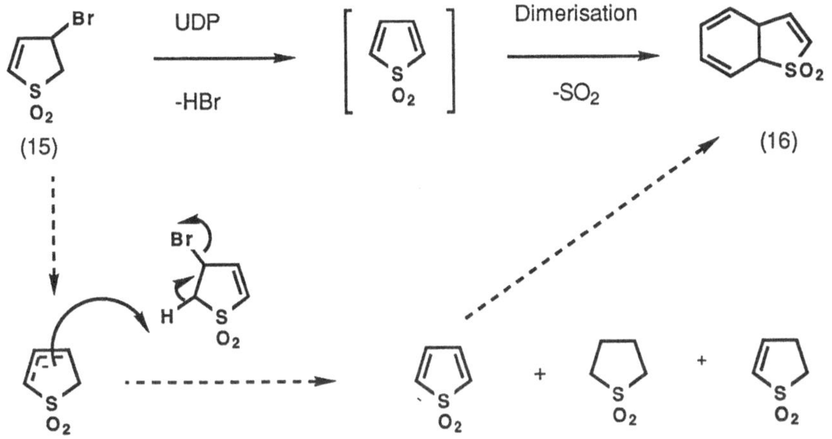

Scheme 47

to (16) was higher than the 50% that would be expected had the second mechanism been in operation. In addition, the 2- and 3-sulpholenes (17) and (18) were not observed as byproducts of the reaction, although a control reaction had shown them to be moderately stable under the reaction conditions.

The presence of a methyl or a chloro-substituent at C3 did not alter the course of the reaction. However, the 3,4-dimethyl derivative (19) gave the thiophene-S,S-dioxide (20), which will react further with UDP to give a mixture of (21) and (22) in a ratio of 10:1 (Scheme 48), accompanied by some of

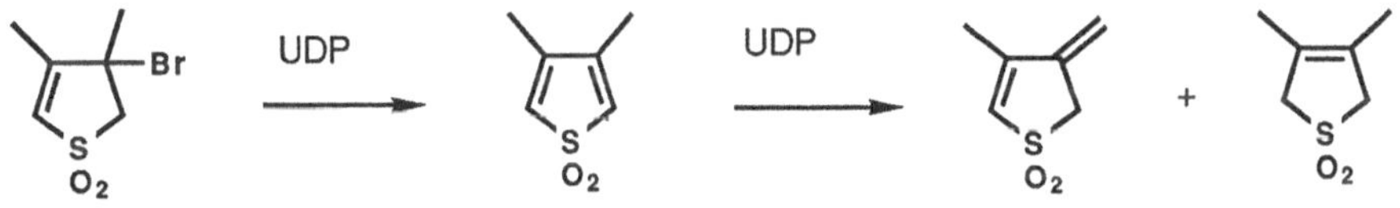

Scheme 48

the unreacted starting material. In fact, deprotonation was favoured over debromination in all but one case. Hence, 2,3-dibromosulpholane (23) gives the sulpholene (24) (Scheme 49). This is presumably due to the presence of the C2 methyl group which blocks dehydrobromination. The selectivity of the process contrasts with the reaction of bromo-sulpholenes in the presence of Ag/Zn or magnesium, when cleavage of the C—Br bond predominates [113]. Hence, UDP appears to exhibit a degree of chemoselectivity in its reaction with a variety of simple sulpholenes and related compounds. The

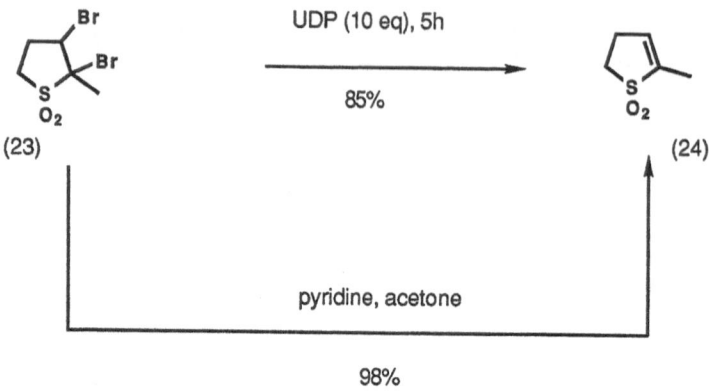

Scheme 49

overall results of these studies can be summarised as shown below (Table 5):

Table 5

Sulpholene		Product	Reference
	1. UDP 2. MeI	—SO$_2$Me	[106]
	1. UDP 2. MeI	—SO$_2$Me	[110]
	UDP		[110]
	UDP		[111]
	UDP		[111]

56

A number of aromatic anions have been generated by sonolysis of mixtures of naphthalene [114, 115], anthracene, biphenyl [115] and 5,6-benzoquinoline [116] with sodium and lithium. Lithium naphthalide (LN), formed as a solution in ethereal solvent, can be transferred easily via syringe and has been used as an effective reducing agent in a number of cases. For example, we have used this methodology to prepare both sodium and lithium naphthalenide which were both shown to be effective agents for the reductive desulphonylation of a number of intermediates that were subsequently used in the synthesis of spiroketals [117] (Scheme 50). In contrast, the thermal reaction is unpredictable and necessitates use of a large excess of the reagent.

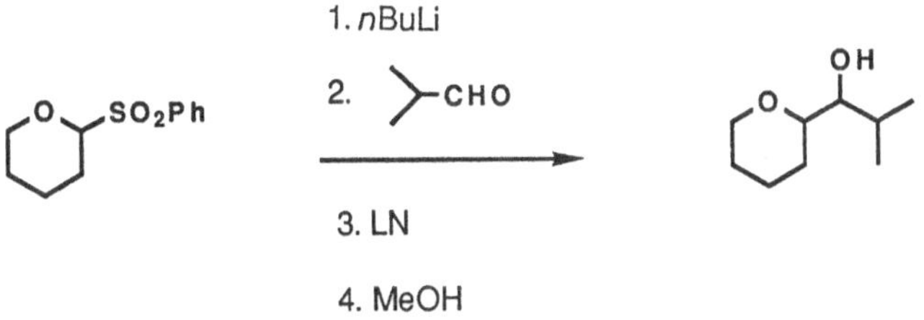

Scheme 50

Solutions of LN in non-ethereal solvent cannot be prepared by thermal means, but the reaction can be brought to completion within 2 h using ultrasound. Solutions of the reagent in TMEDA or TMDAP/benzene were used to promote the dimerisation of isoprene [118] or its reaction with secondary amines [119]. However, the advantages over using a standard solution of the reagent in THF appear to be slight.

In the preparation of KAPA and NAPA, reagents used for the isomerisation of internal acetylenes, sonication of propanediamine with potassium or sodium metal in the presence of a catalytic amount of ferric nitrate generates the reagent in high yield within 10 min [120] (Scheme 51).

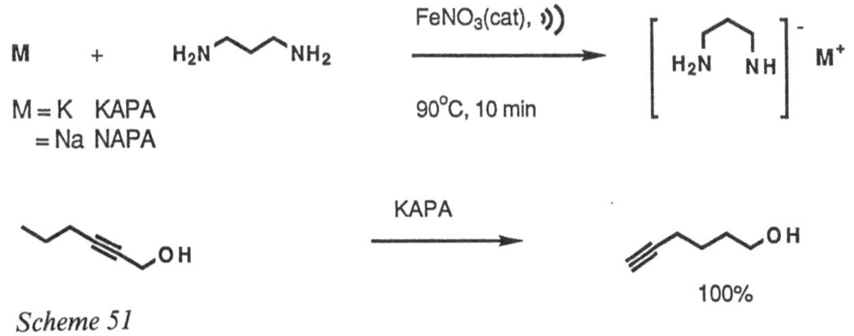

Scheme 51

Previous preparations advocated use of the hazardous and expensive potassium hydride or amide, which must be freshly generated each time. This development is clearly the method of choice.

Ultrasonically dispersed lithium, magnesium and sodium have all been used in the synthesis of allenes and cyclopropylidenes from dihalocyclopropanes [121] (Scheme 52).

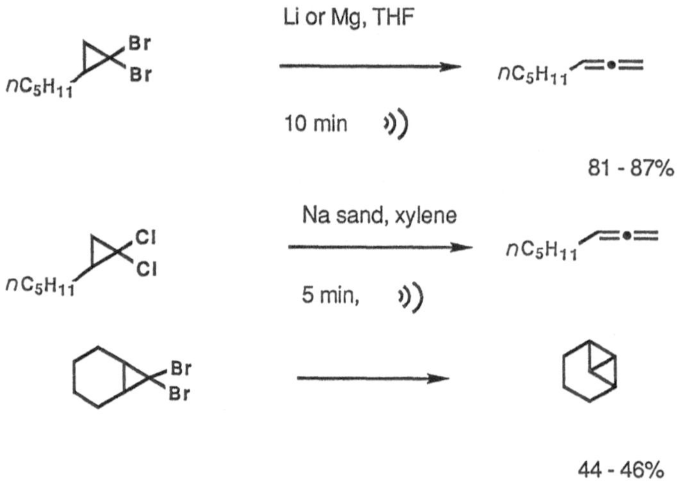

Scheme 52

Yields were high despite the fact that these reactions were all carried out in air using commercial THF which had simply been dried over potassium hydroxide.

Generation of the cyclopropylidene from 7,7-dibromobicyclo-[4.1.0]-heptane in the presence of alkenes gives a variety of novel spirocompounds [122] (Scheme 53).

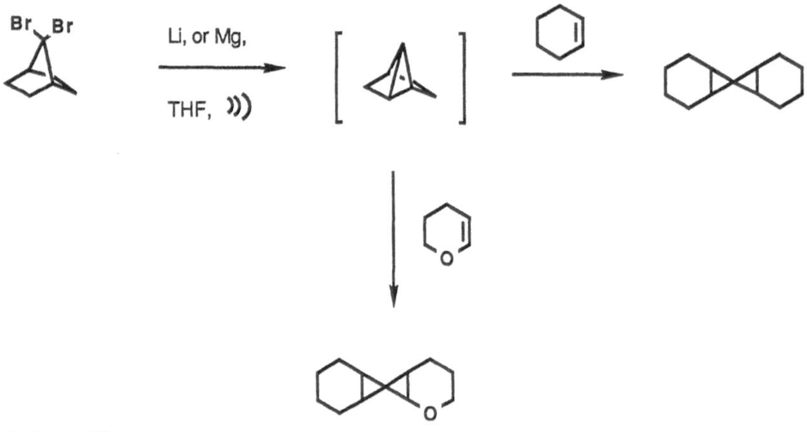

Scheme 53

These ultrasonic reactions are all extremely fast and transform these colloidal dispersions of alkali metals from esoteric to readily accessible reagents. Hopefully, these examples will prompt further exploration of their potential.

8 Organozinc Reagents

8.1 Preparation of Organozincs

Alkylzincs were amongst the first organometallic compounds to be used in organic synthesis. However, they were rapidly superceded by the Grignard reagents which were easier to handle and more reactive. Nevertheless, they are still commonly employed in both the Reformatskii and Simmons-Smith cyclopropanation reactions and recent work on allylzinc bromides suggests that they are potentially useful precursors of olefinic compounds. Significant improvements in the generation of organozincs for these purposes have been made by employing ultrasound and a number of highly reactive species have been shown to be readily available. Luche and coworkers have published a series of papers on the conjugate addition of dialkyl and diarylzinc compounds, prepared using ultrasound, to α-enones. These latest results increase the scope for use of organozincs by demonstrating the analogies between their behaviour and the commonly used organocopper reagents.

8.2 The Reformatskii Reaction

Reaction of a carbonyl compound, commonly an aldehyde or a ketone, with an α-halogenoester in the presence of zinc gives β-hydroxyesters which can be further transformed to give α,β or β,γ-unsaturated esters or, with further elaboration, the dihomologue of the original carbonyl compound. This reaction has found wide application in organic synthesis [123].

Modification of the original procedure has resulted in significant improvements in the yields available from this reaction. These include use of freshly prepared zinc powder [124], a heated column of zinc dust [125] and a trimethylborate/tetrahydrofuran solvent system [126].

In 1982 Han and Boudjouk [127] demonstrated that sonication of a mixture of commercial zinc powder, aldehydes or ketones and α-bromoethylacetate gave extremely high yields of hydroxyesters (**25**) within 5–30 minutes in a one-pot process. Furthermore none of the side products, which result from dehydration of the product or dimerisation of the bromoester, were isolated (Scheme 54).

This methodology was also used to provide a synthesis of monocyclic β-lactams (**26**) in 70–90% yield [128] from a variety of imines (Scheme 54). However, replacing the ethyl bromoacetate by the corresponding 2-bromo-3-phenylpropionate did not give any β-lactam. Nevertheless,

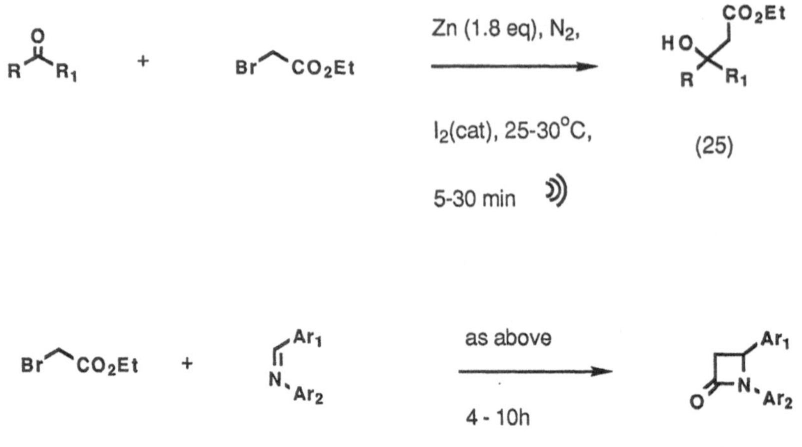

(25)

(26)

Scheme 54

a slight modification of the published procedure allowed synthesis of a variety of monocyclic β-lactams bearing both alkyl and aryl substituents at C3 [129] via a one-pot reaction (Table 6).

Table 6

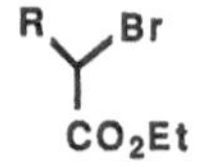 R⎯Br / CO$_2$Et	% Yield of β-lactams 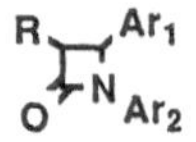	cis : trans
R= Me	85	89 : 11
iPr	92	97 : 3
Bz	87	96 : 4
Ph	88	0 : 100

Interestingly, it appears that the nature of the substituent determines the stereochemical outcome of the reaction, since ethyl 2-bromo-2-phenyl-propionate itself gave the *trans* isomer exclusively. In contrast, reaction in the cases where R was an alkyl group gave the *cis* isomers as the predominant product. This was reasoned to arise from the nature of the zinc enolate formed.

Han and Boudjouk discovered that both I$_2$ and KI could be used to initiate the reaction [127]. Both sets of authors [127, 128] expressed surprise that dioxane proved the optimum solvent in view of its tendency to promote enolisation. However, following an extensive investigation of the reaction conditions, Han and Boudjouk concluded that the reactivity of the system was dominated by the interaction between the zinc surface and the iodine. Oguni's one-pot reaction does not require addition of a discrete initiating agent. However, this role is probably fulfilled by the α-bromopropionate itself.

These conclusions are in agreement with observations that metal powders can be activated by halide salts [130], and evidence produced by electron microscopy of the nature of the surface of manganese activated by iodine in the presence and absence of ultrasound [131] (Scheme 55).

Scheme 55

These show that the number of initiation sites generated on the surface of the metal by a short period of sonication far exceed those created by simply stirring the reagents together for several hours. Furthermore, this investigation into the coupling of allylic iodides in the presence of activated manganese conclusively demonstrated that this was the major role of the ultrasound since, once the initiation process was complete, the rate of the ongoing reaction was unchanged by the presence or absence of ultrasound.

An extension of this procedure to activate zinc allows preparation of the reduced penicillinate esters from 6-aminopenicillanic acid by way of the organozinc derivative [132]. There are a number of existing methods for effecting this transformation. For example, the C—Br bond can be reduced using tri-*n*-butyltin hydride [133], or by palladium-catalysed hydrogenation [134]. However, both of these methods suffer from a number of disadvantages. In the first case, removal of tin residues from the final product is not trivial. Secondly, both reagents are relatively expensive and this is compounded by the presence of the sulphide group which necessitates use of almost quantitative amounts of palladium if hydrogenation is employed.

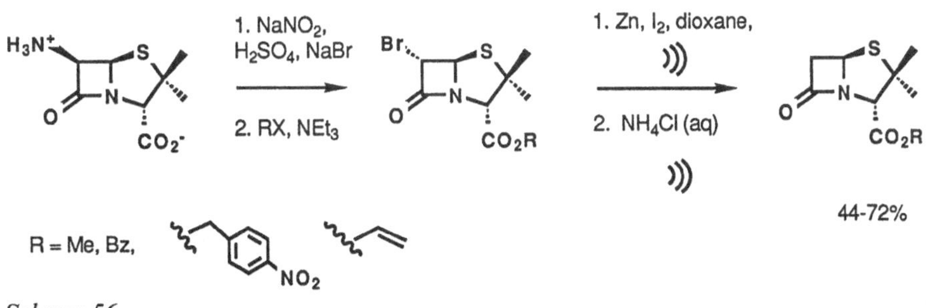

Scheme 56

Brennan and Hussain have shown that formation of the organozinc intermediate occurs readily under identical conditions to those described for the Reformatskii reaction in anhydrous dioxane [132]. In situ hydrolysis then gives the penicillanate esters in moderate yield (Scheme 56). This process was sucessful in the presence of a number of common carboxyl-protecting groups; the one exception being the case where R = *p*-nitrobenzyl when unchanged starting material was re-isolated. In addition, the procedure was equally successful in the case of the equivalent sulphones and sulphoxides.

8.3 Generation of Allylzinc Reagents

Allylic, benzylic and propargylic bromides will also react with zinc to give the corresponding organozinc bromides. Reaction with both C—C and C—N multiple bonds has already been shown to provide a potentially useful route to some ȯlefinic compounds [135]. However, the reaction with terminal acetylenes produces mixtures of mono- and bis-addition products (27) and (28) in low yields [136] (Scheme 57).

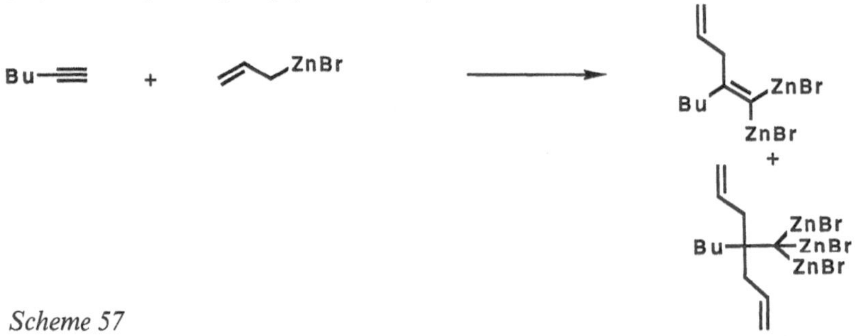

Scheme 57

The synthetic utility of this reaction was greatly increased when Knochel and Normant [137] showed that sonicating the reaction mixture allowed mono-addition to predominate giving good yields of highly functionalised dienes (29). Further elaboration gave a variety of six and seven membered carbo- and heterocycles (Scheme 58).

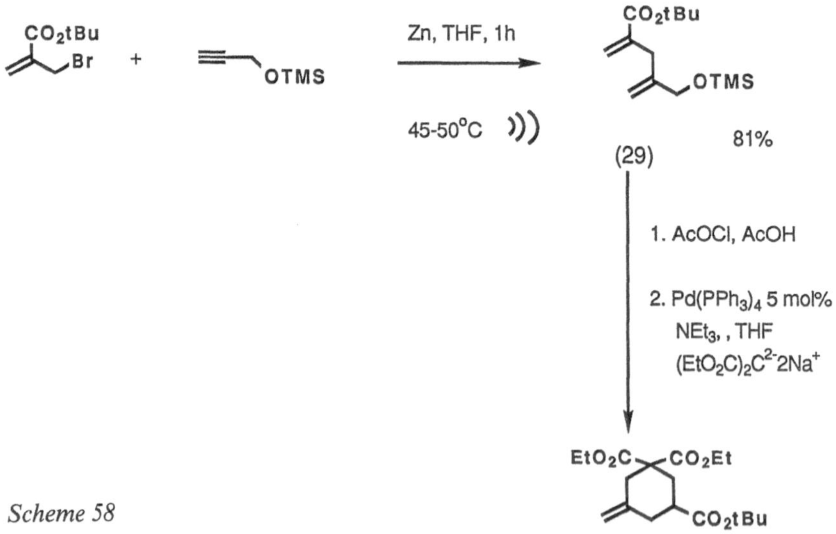

Scheme 58

8.4 Zinc Mediated Cyclopropanation

Reaction of alkenes with diiodomethane in the presence of a zinc/copper couple produces cyclopropanes by methylene transfer from an organozinc species. This is the established method of synthesising cyclopropanes. However, the procedure requires activation of the zinc by use of Zn/Cu, Zn/Ag couples, I_2 or lithium metal. The first of these is most commonly used, but the sensitivity of the system to air makes the results irreproducible and erratic. Furthermore, the reaction which follows has an induction period of indeterminate length and can be extremely violent. Both these factors contribute to making large scale reactions fickle and dangerous.

In 1982 Repic and Vogt [138] discovered that chemical activation was unnecessary in the presence of ultrasound and yields of cyclopropanated olefins were considerably higher than could be obtained using conventional methods. The reaction was carried out under nitrogen in refluxing dimethoxyethane, taking care to ensure that all traces of iodine were removed from the diiodomethane prior to its use. Under these conditions reaction was instantaneous and was found to proceed smoothly and rapidly with extremely crude forms of zinc, e.g. rods, foils and mossy zinc. Examination of the reaction with methyl oleate in the presence and absence of ultrasound clearly demonstrates the rate increase obtained by sonication of the reaction mixture; this is shown in Fig. 22. Furthermore, switching off the ultrasound source reduces the rate of reaction to that of the stirred reaction.

The same reaction was scaled up and run in a 50-gallon bath at 100 °C [139]. The reaction gave 0.5 kg (82%) of the cyclopropane after 2.25 h. In this case the zinc was cast as two 800 g lumps (using 125-ml conical flasks as moulds!). This increased the degree of control possible as the surface

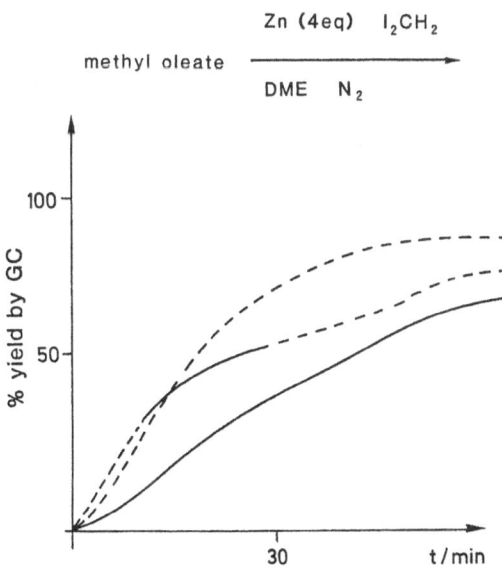

Fig. 22. Percentage yield of cyclopropanated material as a function of time for the reaction of methyl oleate and zinc/diiodomethane in dimethoxyethane showing the differences in rate of reaction between the sonicated (--------) and stirred (———) reactions

of the zinc available is not as large as that presented by powdered zinc. Hence, the exotherm is more evenly distributed and the lumps of zinc can simply be removed from the reaction mixture once reaction is complete.

Reaction of the organozinc reagent with β,γ-unsaturated ketones gives the relevant furan as shown in the case of lipid (**30**). The authors examined the reaction using both cadmium and copper in place of the zinc and found that whilst zinc gave the highest yield for cyclopropanation adjacent to a carbonyl group, it was surpassed by copper in the case of the equivalent allylic alcohol (Scheme 59).

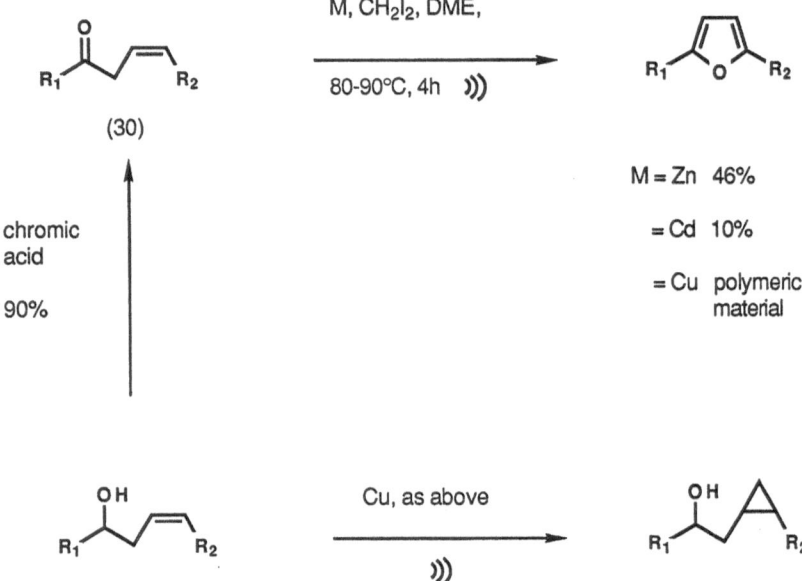

Scheme 59

For reasons previously explained, it seems unlikely that the high temperatures reported are necessary for generation of the organozinc. This view is supported by evidence that aldehydes can be methylenated using the same organozinc prepared in THF solution at room temperature [140]. Unoptimised yields of olefin (**31**) were between 30 and 70% within 20–120 min. Yields obtained using ketonic substrates were much lower — although comparable to those obtained using an activated Zn/Cu couple. This is of no

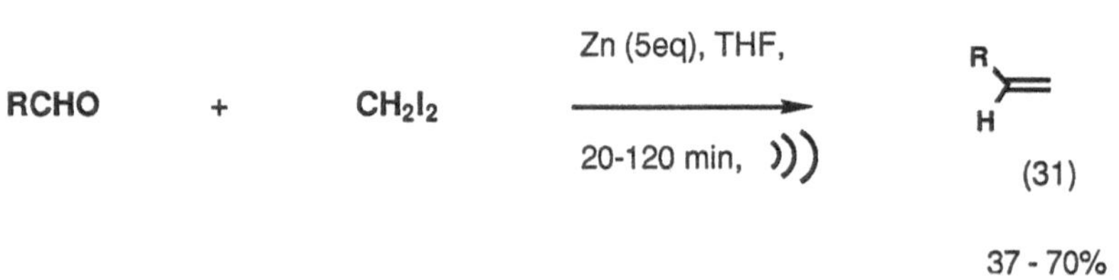

RCHO   +   CH₂I₂ → (31)

Zn (5eq), THF,

20-120 min,)))

37 - 70%

Scheme 60

great consequence as complimentary methodology for the methylenation of ketones already exists.

Recently published work [141] showed that the less reactive dibromomethane could be used for cyclopropanation although the prior activation of the zinc was necessary. The reaction has a long induction period and yields are somewhat lower than those reported using diiodomethane.

This effect is particularly noticeable in the case of acyclic alkenes, although yields could probably be improved if the reaction was carried out at lower temperature. Nevertheless, the authors suggest that the current disparity between the prices of the two dihalomethanes argues cogently in favour of its use — particularly in the preparation of cyclopropanes from readily available starting materials (Scheme 61).

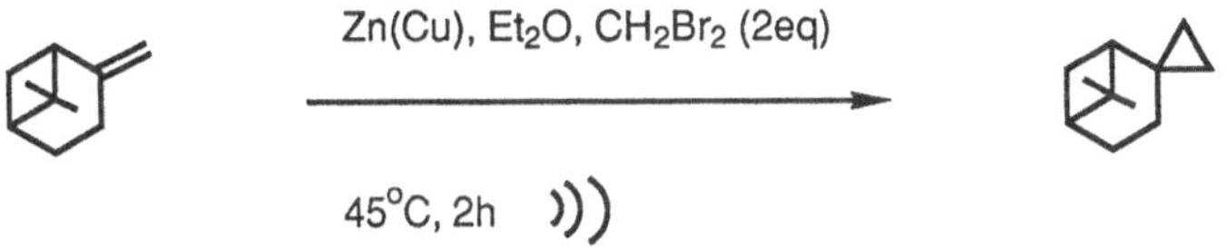

Zn(Cu), Et$_2$O, CH$_2$Br$_2$ (2eq)

45°C, 2h)))

Scheme 61

8.5 Formation of Alkylzinc Species

In 1977 Isobe et al. reported that R$_3$ZnLi $\cdot$ 2 LiCl species would add conjugatively to α-enones in a 1,4 fashion. Transfer of the alkyl groups proceeded in high yield, although neither phenyl nor acetylenic substrates could be persuaded to react at all [142].

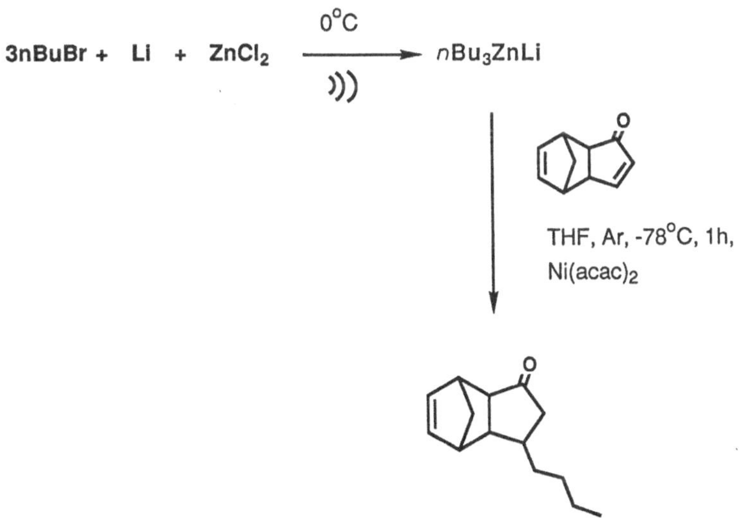

0°C

3nBuBr + Li + ZnCl$_2$ ———→ nBu$_3$ZnLi

)))

THF, Ar, -78°C, 1h,
Ni(acac)$_2$

Scheme 62

In 1983 Luche's group discovered that sonolysis of a mixture of aryl bromide, lithium wire and zinc bromide in an ethereal solvent at 0 °C resulted in the formation of organozinc species without the need for use of pre-formed air sensitive alkyllithiums.

These added conjugatively to α-enones in the presence of Ni(acac)$_2$ (Scheme 62). Side reactions, e.g. Wurtz coupling, were minimal and the reaction gave good yields of the 1,4-addition products — even in the case of β,β-disubstituted enones [143]. This was essentially a one-pot reaction and complemented the results of Isobe's original report [142].

However, the addition of alkyl and vinyl bromides was reported to be "unoptimised", that is to say that results obtained using a Sonoclean (40 kHz, 96 Wl^{-1}) bath were both erratic and irreproducible.

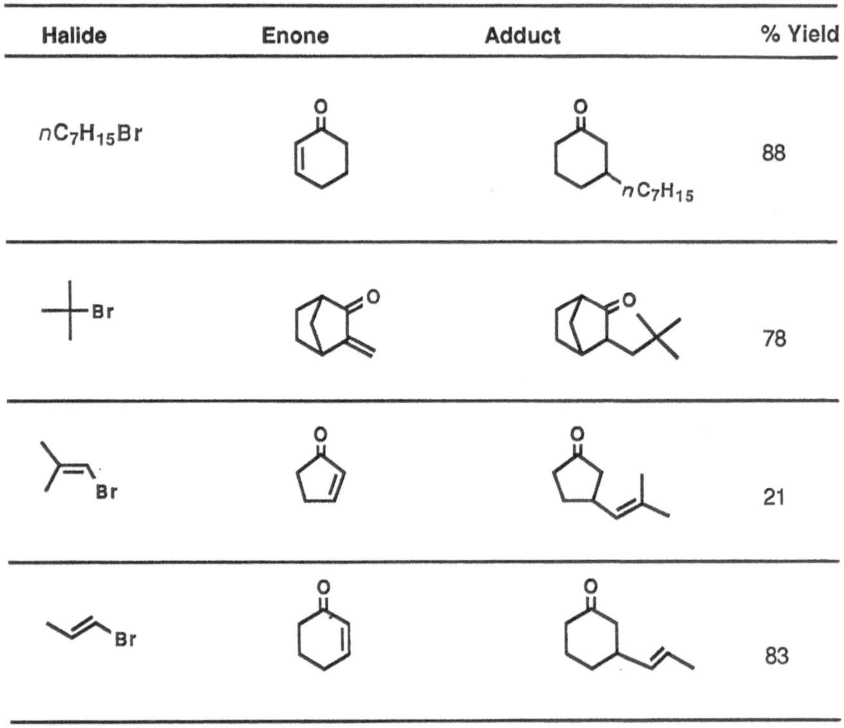

Halide	Enone	Adduct	% Yield
$nC_7H_{15}Br$			88
			78
			21
			83

Scheme 63

This problem was overcome by use of a cell-disruptor type ultrasound generator. This modification permitted much greater control of the sonication conditions and allowed addition of 1°, 2°, 3° alkyl and allylic groups in essentially quantitative yield within 10–20 min [144] (Scheme 63).

This development allows analogies to be drawn between the synthetic utility of the organozinc reagents and the organocuprates commonly used for such transformations. The thermal stability of the organozincs compares well against the sensitivity of the organocuprates, although the stoichiometry of the reaction still requires the use of two equivalents of alkyl bromide. Recent attempts to overcome this problem by use of RZnX species failed to result in any significant transfer of the alkyl group to the α-enone [145] (Scheme 64).

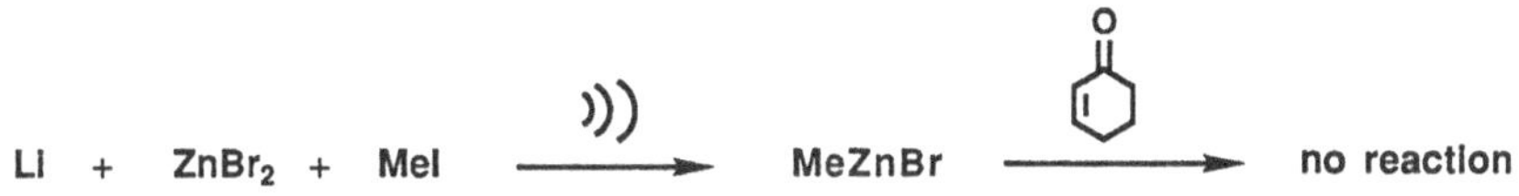

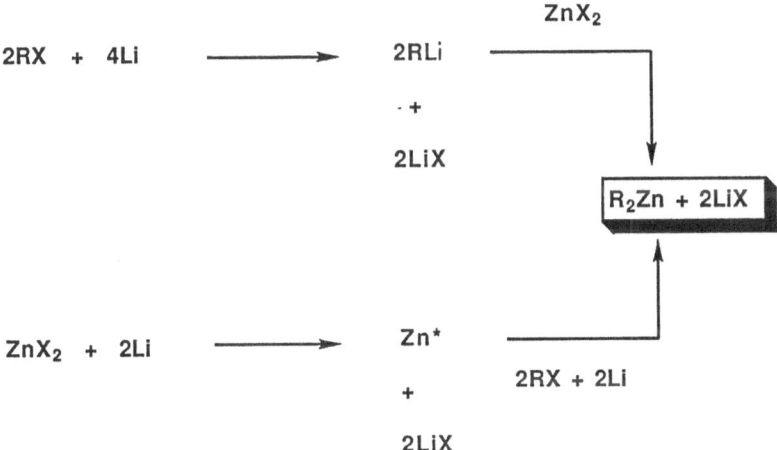

Li + ZnBr$_2$ + MeI $\xrightarrow{\text{)))}}$ MeZnBr $\longrightarrow$ no reaction

Scheme 64

In their original communication [143] Luche et al. had assumed that the active species was a diarylzinc on the basis of the stoichiometry of the reagents used. They suggested that such species could be generated via one of two possible pathways. The primary process could either be envisaged in terms of formation of an aryllithium followed by transmetallation with zinc bromide to give the organozinc, or generation of activated zinc resulting from reduction of the zinc halide by the lithium. Reaction with the aryl halide in situ would then lead to formation of the arylzinc (Scheme 65).

ZnX$_2$

2RX + 4Li $\longrightarrow$ 2RLi

· +

2LiX

R$_2$Zn + 2LiX

ZnX$_2$ + 2Li $\longrightarrow$ Zn*

+

2LiX

2RX + 2Li

Scheme 65

The second pathway is analogous to the reaction of an activated metal, similar to those generated using Riecke's methodology, with alkyl halides. However, Luche reports that pre-activation of the zinc gave poor yields of organozinc under sonochemical conditions.

It should also be noted that a large excess of lithium was used in all cases. Luche has not published any experimental data to support one or other mechanism but he clearly favours the transmetallation argument [145].

In the same paper Luche et al. have also shown that this methodology can be applied to the formation of mixed organozincs in the manner shown below (Scheme 66).

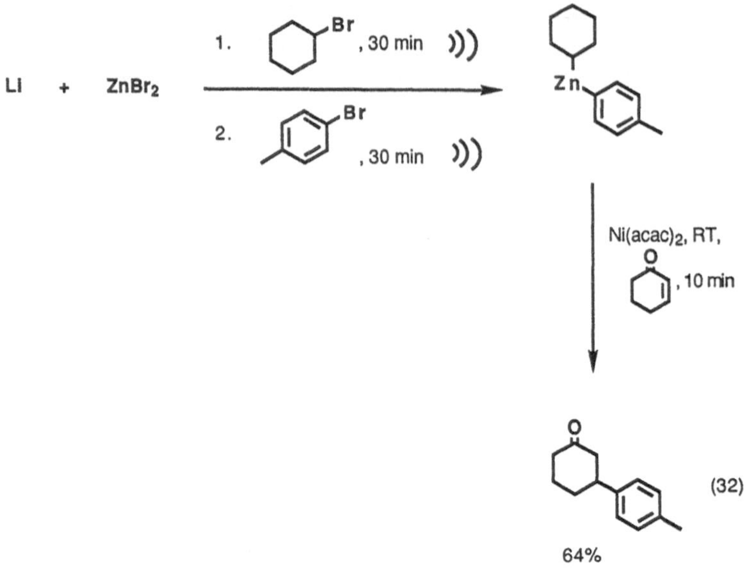

Scheme 66

The only product isolated (**32**) arises from transfer of the aryl group to the enone. This complements the behaviour of mixed organocuprates which will transfer unsaturated alkyl groups in preference to saturated ligands.

In a recent publication Petrier and Luche [62] have published some interesting work concerning the reaction of allylzinc and tin species with aldehydes in aqueous media. Allylzinc reagents had been found to add to α,β-unsaturated carbonyl groups in a 1,2-rather than the usually observed 1,4-fashion (Scheme 67). The erratic results obtained suggested that reaction might be occurring on hydrolysis of the reaction mixture.

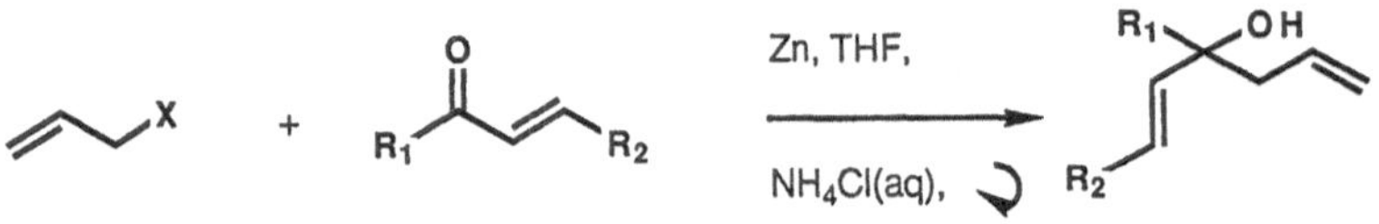

Scheme 67

This was later confirmed when it was shown that reaction of a carbonyl compound with an unidentified species, formed by reaction of the allylic halide with zinc, could be effected in a mixture of THF and saturated aqueous ammonium chloride solution. Sonication was not necessary and the product obtained was the expected homoallylic alcohol. Spontaneous reaction also occurred in water, but at a greatly reduced rate and in these cases reaction was greatly accelerated in the presence of ultrasound. Furthermore, metallic tin could also be used in place of zinc (Schemes 7, 8, and 68).

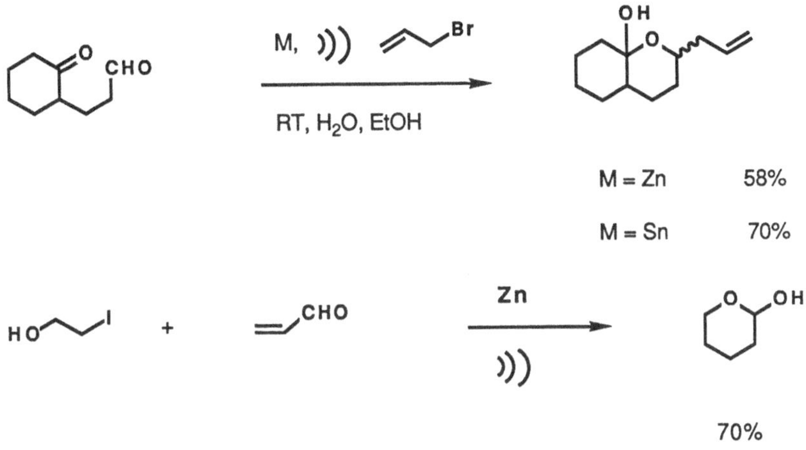

M = Zn   58%

M = Sn   70%

70%

Scheme 68

Yields for both the stirred and sonicated reactions were of the same order of magnitude, although the zinc reagents appeared to be less sensitive to steric crowding than the corresponding tin analogues. Competitive reaction between aldehydes and ketones favoured addition to aldehydes and even unactivated alkyl halides could be persuaded to add conjugatively to α-enones and enals if the zinc was activated by a copper couple [146]. The specificity of this last reaction is particularly interesting given that the intermediate aldehyde enols are generally highly reactive and polymerise readily under standard reaction conditions.

Reactions have been reported in aqueous solutions of ethanol, acetone, THF, and pyridine, and even in pure water itself. The mechanism for this is is not understood although a plausible suggestion would be that the reaction proceeds via a radical pathway, which would be promoted by the presence of water, rather than via any discrete organometallic species.

Organozincs generated under similar conditions have been shown to add to optically active vinyl phosphine oxides with retention of configuration at phosphorus [63] (Scheme 8). In this case a 1:1 zinc/copper couple was employed to activate the zinc. The one-pot reaction proceeded best with tertiary or secondary alkyl halides. The resulting phosphine oxides are used as chiral ligands for catalysts in asymmetric synthesis and were previously only available by reaction with the analogous cuprate [147] since conjugate addition of simple alkyl halides normally results in polymerisation.

As yet this methodology remains unexploited. In addition the comparative hardness and rapid, facile preparation of these reagents should make this methodology of great interest to synthetic organic chemists. Luche and co-workers have clearly demonstrated the scope of organozincs as alternatives to organocuprates.

8.6 Zinc Mediated Perfluoroalkylation

Perfluoroalkylmetals are not commonly used in synthesis — presumably because of their low stability. The driving force behind this is the elimination of the metal halide leaving the perfluoroolefin.

Kitazume and Ishikawa have published a number of communications on the regiospecific zinc mediated perfluoroalkylation of a variety of substrates [148–151]. A compilation of these results appeared in 1985 [152] (Scheme 71). Secondary alcohols are available in moderate yield by reaction of perfluoroalkyl halides (R_FX) with aldehydes as the result of a modified Barbier-type reaction [148–152]. However, reaction with ketones gave low yields of the expected tertiary alcohols, although the yields could be doubled by reaction in the presence of a bis[π-cyclopentadienyl]titanium(II) catalyst, which was prepared in situ by reduction of the dichloride with zinc in the presence of ultrasound.

In contrast it should be noted that yields in the analogous lithium mediated reaction were consistently higher without the need for use of any additional catalysis [83]. Both vinylic, allylic and aryl perfluoroalkylzinc compounds [149, 152] could also be generated in the presence of palladium catalysts. The regiospecificity of the reaction was reported to be better than 95 % although yields of the alkylated product were only moderately good. The addition of perfluoroalkylzincs to alkynes and dienes was regio-but not stereo-specific [151, 152] and, as in all cases, yields of alkylated products were consistently higher using alkyl iodides than with the equivalent alkyl bromides.

A brief investigation using chiral auxiliaries such as Ender's RAMP and SAMP, provided a route to α-disubstituted ketones (32) with a moderate degree of chiral induction [152] (Scheme 69).

The only other example of a stereospecific perfluoroalkylation reaction also used Kitazume's methodology. Using a chromium tricarbonyl moiety as the chiral auxiliary gave high yields of benzylic alcohol (33) via a modified Barbier reaction [153]. However, the degree of asymmetric induction using unbranched perfluoroalkyl halides was low, although a marginal increase was seen when using branched halides. The chromium tricarbonyl

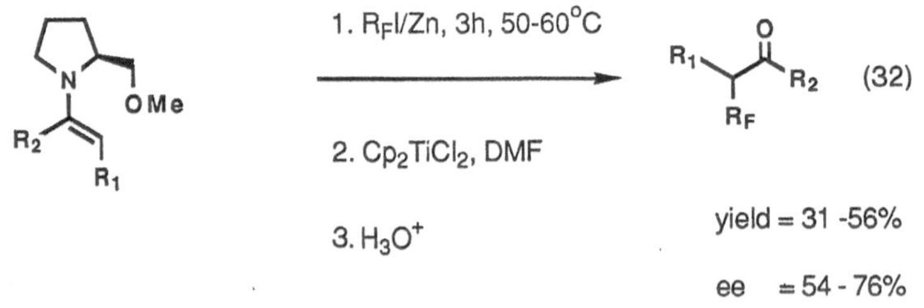

1. R_FI/Zn, 3h, 50-60°C

2. Cp_2TiCl_2, DMF

3. H_3O^+

(32)

yield = 31 -56%

ee ≅ 54 - 76%

Scheme 69

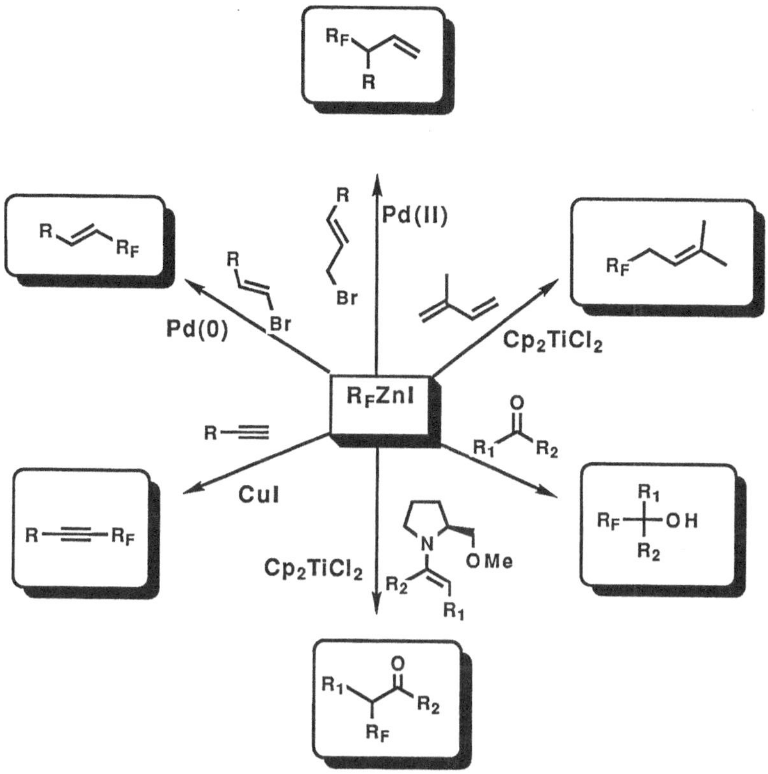

Scheme 70

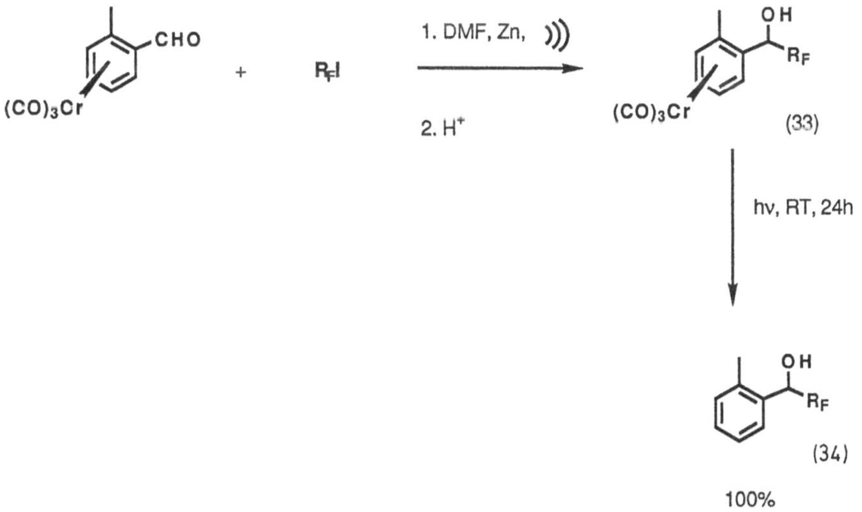

Scheme 71

moiety is readily removed by photolysis and yields of the alcohol (34) are virtually quantitative (Scheme 71).

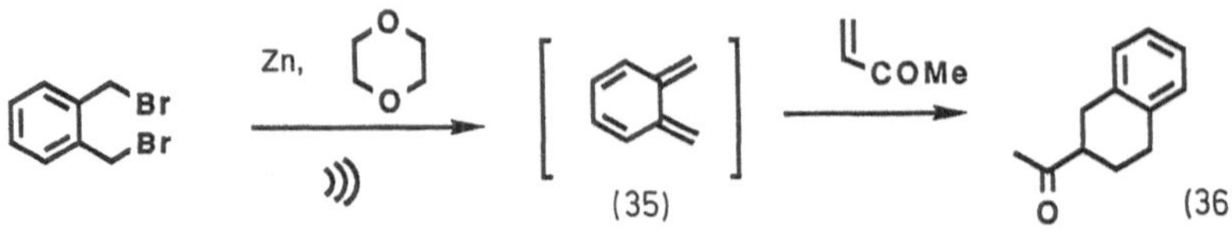

Scheme 72

8.7 Zinc-mediated preparation of Reactive Intermediates

Ultrasound has been exploited in the zinc mediated generation of a number of reactive intermediates. Most of these are only normally attainable in very low yield as a result of the forcing conditions necessary for their generation. On the other hand Han and Boudjouk showed that ortho-xylylene (35) was readily available from o-dibromoxylene. Trapping the intermediate tetraene with methyl vinyl ketone gave an 87% yield of the product (36) arising from Diels-Alder reaction with the exocyclic diene [154] (Scheme 72).

20 - 30%

Scheme 73

This methodology has been exploited in the synthesis of a number of tri- and tetra-cyclic systems related to the fungal metabolite altersolanol A and the anthracyclinone β-rhodomycinone. As expected, cycloaddition occurred from the β-face of the dieneophile providing a stereospecific, if low yielding, route to these systems [155] (Scheme 73).

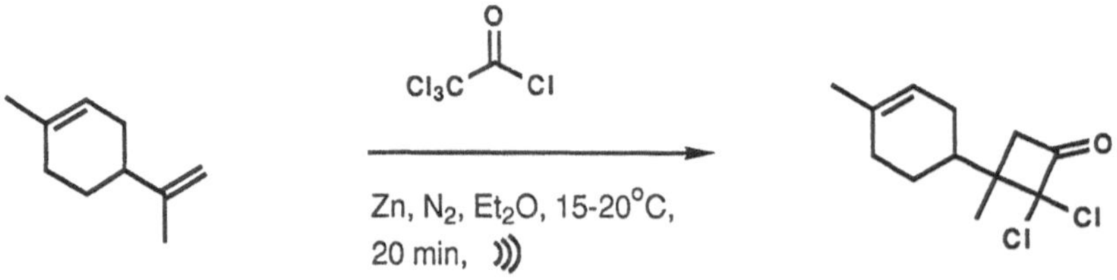

Zn, N$_2$, Et$_2$O, 15-20°C, 20 min,)))

80%

Scheme 74

The potential for introduction of chiral centres using a Diels-Alder approach makes this a particularly attractive route to benzo-fused cyclo-hexanes. However, the chemistry of orthoxylylenes is relatively unexplored as previous routes to this elusive intermediate involved the use of reagents such as disodium tetracarbonylferrate. This reagent is extremely intolerant of other functional groups, mainly as a result of its high basicity and the yields obtained are only a fraction of those available using the zinc/ultrasound route.

Dichloroketene is another useful reagent which can only be generated in situ. This is normally achieved by treatment of Cl_3CCOCl with activated zinc. Nevertheless, despite a number of modifications, reactions require a large excess of olefin and cycloadditions can take long periods of time — overnight is unexceptional. By contrast, sonication allows the reaction to be carried out in a fraction of the time. Furthermore, the reaction can be carried out using commercial zinc powder and the authors report that reaction is generally complete in the time required for addition of the chloride [156].

The reagent generated in this way shows a high degree of selectivity, as illustrated above (Scheme 74). Furthermore, reaction with 1,5-cycloocta-diene produces a 70% yield of the monoadduct uncontaminated by any of the diadduct (Scheme 75). The equivalent reaction using a zinc/copper couple produces mixtures of the mono- and di-adducts [157].

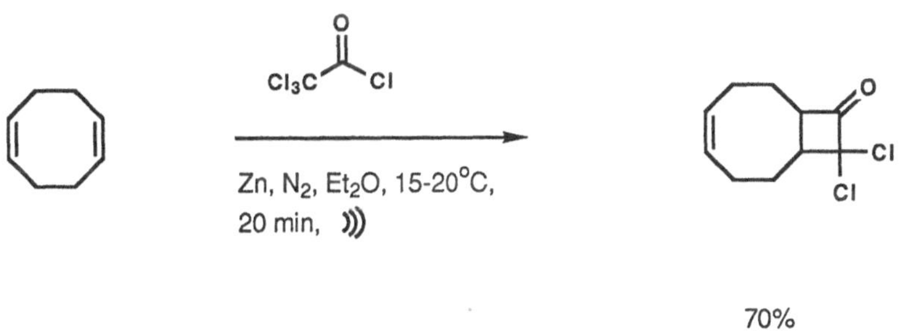

Zn, N_2, Et_2O, 15-20°C, 20 min,))

70%

Scheme 75

Preparation of bicyclic systems via the reductive coupling of 1,3-dienes with α,α′-dibromoketones to give bicyclic systems has been known for some time. Hoffmann discovered that moderate yields of coupled product were available using a zinc/copper couple. The reaction took between 12–80 h [158]; however, yields could be significantly improved by use of a trimethyl-silylchloride mediator [159]. Complementary work by Noyori [160] had already shown that use of diiron-nonacarbonyl led to significant increases in yield despite the obvious disadvantages of its expense and toxicity. Hoffmann has recently examined the effects of ultrasound on his original reaction [161].

73

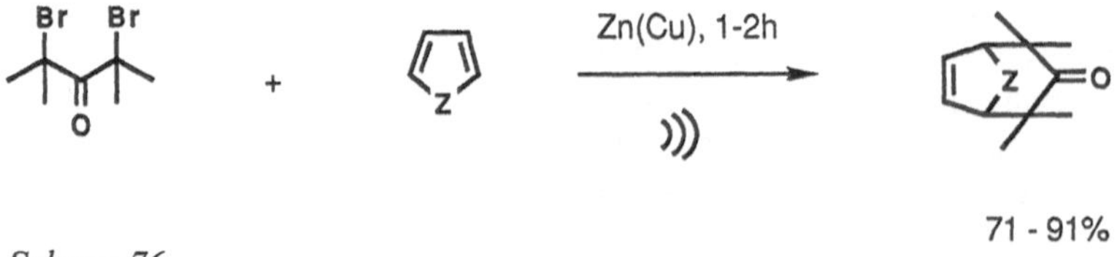

71 - 91%

Scheme 76

He reports that the reaction proceeded smoothly even in the absence of the silyl chloride, although reaction times could be reduced to less than 30 min in its presence (Scheme 76). In fact, the highest yields were obtained using sterically hindered tetrasubstituted ketones and the method appears to be indispensible in such cases. In addition, ultrasound was reported to promote the diiron-nonacarbonyl mediated reaction, although reaction times were somewhat longer.

Extensive investigations into the alkylation of α,α'-dibromoketones using mercury dispersions have been carried out by Fry and co-workers [162–169]. Comparison with the equivalent electrochemical route [162] showed that reduction with ultrasonically dispersed mercury gave a higher degree of regioselectivity. Alkylation proceeded smoothly with a variety of nucleophiles and condensation with ketones produced a one-pot synthesis of 4-isopropylidene-1,3-dioxalans (**38**) (Scheme 77). Reaction is thought to proceed through the 2-hydroxyallyl cation (**37**) and mono-substitution of the parent ketone predominates — even in the case of sterically undemanding substrates. Fry believes that this is purely a mixing effect and reports that the reaction proceeds equally well using a high speed laboratory homogeniser.

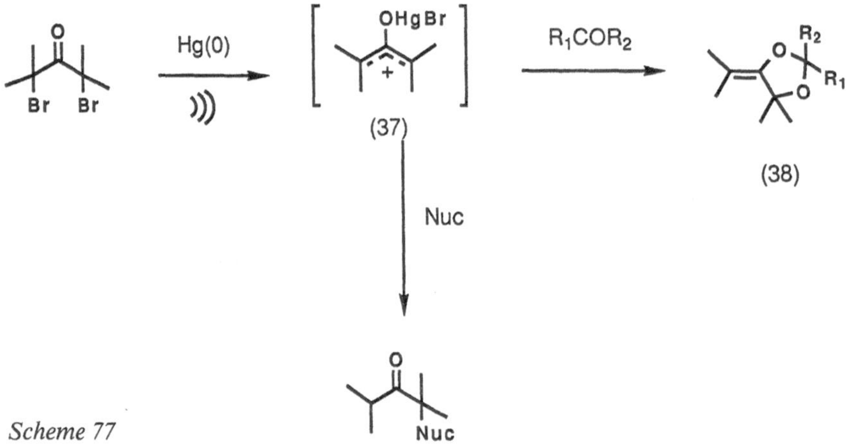

Scheme 77

9 Intercalation Reactions

Suslick and Green have reported the profound effects of ultrasound on the intercalation of various guest molecules into layered inorganic solids [170]. The products of these reactions are of interest in attempts to design new materials with novel optical, electrical and catalytic properties. However, synthesis of such compounds generally requires the use of extremely high temperatures over very long periods of time. In contrast, sonicating the reaction mixture brings the reaction to completion within hours, rather than the days required for the thermal reaction.

Examples given in a later publication [171] detail the preparation of a variety of (guest)$_x$host complexes (Table 7). In each case the sonochemical reaction was carried out by sonicating a mixture of the host solid and guest molecule (2:1) in toluene using a direct immersion probe operating at 20 W/cm^3.

Table 7

(Guest)$_x$host	Reaction conditions	
	Thermal	Sonochemical
1. (Cp$_2$Co)$_{5.25}$ZrS$_2$	50 h/20 °C	2 h/20 °C
2. (RNH$_2$)$_x$TaS$_2$*	50 h/20 °C	15 min/20 °C
3. (pyridine)MoO$_3$	30 days/180 °C	3 days/80 °C

* R = n-Bu, x = 0.46
  R = n-hexyl, x = 0.40

Closer examination of the reaction between n-hexylamine and TaS$_2$ suggested that the increase in intercalation rates was not due to increased mass transport, but to an irreversible change in the host. Electron microscopy clearly shows that the initial effect of sonication is to reduce the size of the solid particles, in this case a reduction from 60–90 μm to 5 μm within the first 15 minutes. Furthermore, the degree of intercalation had almost reached its maximum value at this point, suggesting that this is the major determinant of the reaction time rather than effects related to subsequent surface damage (Plate 3, see p. 16).

10 The Effects of Ultrasound on Enzyme-Catalysed Reactions

The use of biological systems for effecting organic transformations has attracted a great deal of interest over the past few years. The advantages of using enzymes to this end lie in their complete specificity for a particular substrate coupled with the unparallelled degree of stereo- and regioselectivity which can be obtained.

Experimental investigations into the behaviour of the enzymes α-amylase and glucoamylase bound to porous polystyrene show that their activity is raised in the presence of ultrasonic irradiation. The maximum activity increase that could be obtained using an acoustic intensity of $5 \, kW/m^2$ amounted to $>200\%$ [172]. Similar increases have been observed for α-chymotrypsin bound to agarose [173]. Schmidt and co-workers have suggested that this is attributable to a reduction of the thickness of the unstirred diffusion layer around the carrier bodies [172], that is to say that these observations are primarily due to increased mass transport rather than any cavitational effects.

Kyler and co-workers have examined the effects of ultrasound on a suspension of baker's yeast (*Saccharomyces cerevisiae*) as an inexpensive, convenient and direct source of sterol cyclase [174] (Scheme 78). This enabled them to produce sterols from squalene oxide and related substrates on a multigram scale. Previous experiments had relied on the use of microsomal cyclase. However, this process can only be used to convert extremely small quantities of substrate to the desired sterol (i.e. <1–3 mg) and is obviously unsuitable for preparative purposes. In contrast, "ultrasonically stimulated" yeast cells could be used to convert 2,3-oxidosqualene to lanosterol in 42%

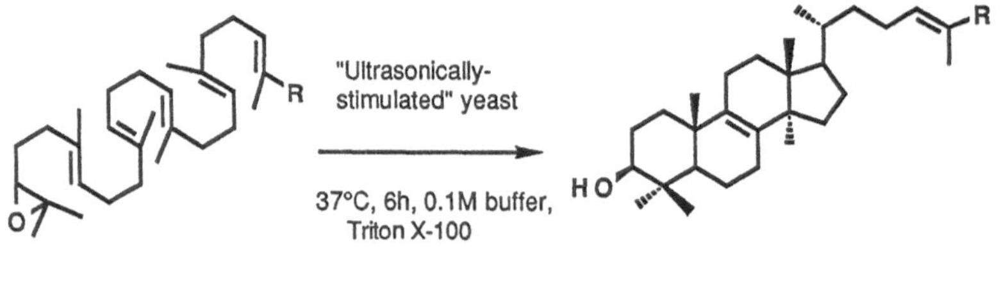

"Ultrasonically-stimulated" yeast

37°C, 6h, 0.1M buffer, Triton X-100

R = Me

= CO₂Me (39)

lanosterol (R = Me)

methyl ganoderate   (R = CO₂Me) (40)

Scheme 78

yield. This actually represents an 84% conversion, given that the (R)-squalene oxide was recovered unchanged from the reaction mixture. In contrast, the maximum conversion that could be obtained using untreated baker's yeast was 19%.

Similarly, the methyl ester of ganoderic acid (**40**) was prepared by cyclization of the squalene (**39**) ($R=CO_2Me$). The authors report that maximum conversion efficiency was obtained after sonication at 0 °C for 2 h at an acoustic intensity of approximately 40 W/cm^2. In contrast, sterol production with a cell-free cyclase system was completely insensitive to ultrasonic irradiation. This suggests that the major effect of ultrasound is to break down the cell wall of the yeast, rather than activating the cyclase or preventing its inhibition. Conversely, the liberation of membrane-associated sterol carrier protein factors cannot be ruled out.

11 Ultrasonic Acceleration of Organic Reactions

Although most of the examples in this review deal with the interaction of metals with organic substrates, the effects observed are not limited to organometallic systems and a number of useful procedures for transformations involving non-metals have been published. For instance, amines can be N-alkylated by alkyl halides in the presence of potassium hydroxide and a phase transfer catalyst [175]. N-benzyl glycine was readily prepared in high yield by sonolysis of the Cu(gly)$_2$ and the relevant chloride at pH 11.0. The process was similarly successful in the case of p-chlorbenzyl chloride, but only gave 5% of the N-arylmethylated product in the case where R = p-Me [176] (Scheme 79).

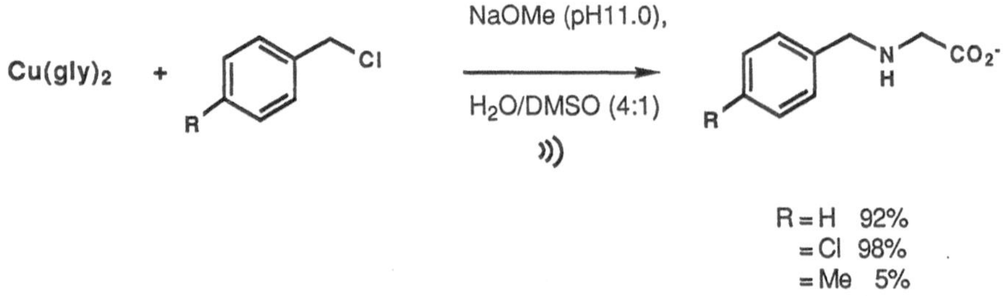

R = H   92%
  = Cl 98%
  = Me   5%

Scheme 79

Similarly, direct displacement of halides by the thiocyanate anion proceeds smoothly to give high yields of the corresponding alkylthiocyanate [177]. Alkylation of isoquinoline using ultrasonically generated dimsyl sodium [178] was also recently reported by workers using methodology developed in 1966 [179] (Scheme 80).

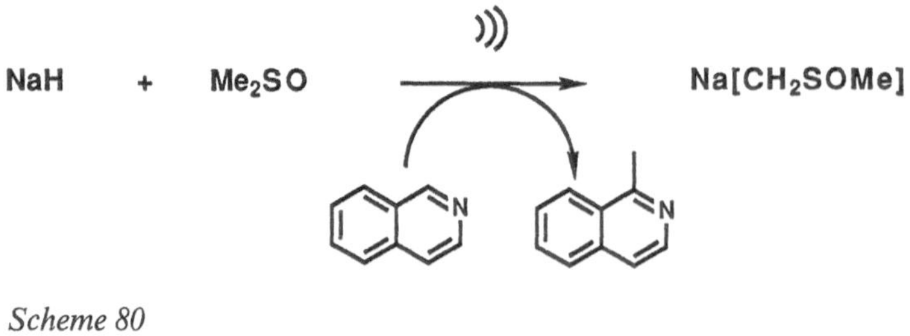

Scheme 80

Alkylation of indene was achieved under milder conditions using aqueous sodium hydroxide as the base. This necessitated use of a phase transfer catalyst (PhCH$_2$NEt$_3$Cl), but yields of 3-alkyl indenes were virtually quantitative [180] (Scheme 81).

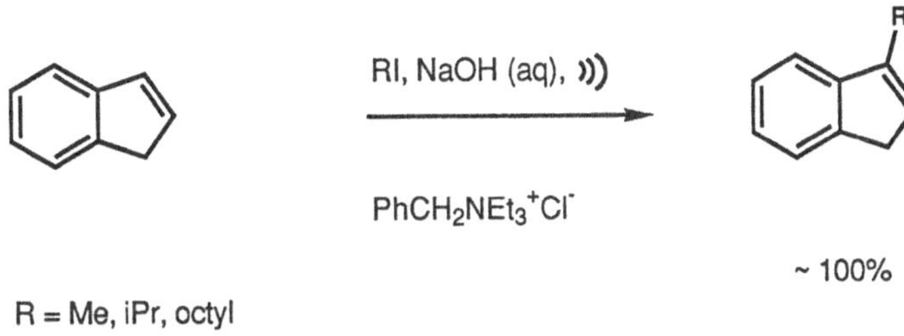

RI, NaOH (aq),)))

PhCH$_2$NEt$_3$$^+Cl^-$

~ 100%

R = Me, iPr, octyl

Scheme 81

N-alkylation of 3-methyl-7-propylxanthine was carried out under apparently neutral conditions by sonicating the reagents in toluene/water (25:1) in the presence of neutral alumina [181]. The system was pre-treated by sonication for 1 h before addition of the alkylating agent (Scheme 82).

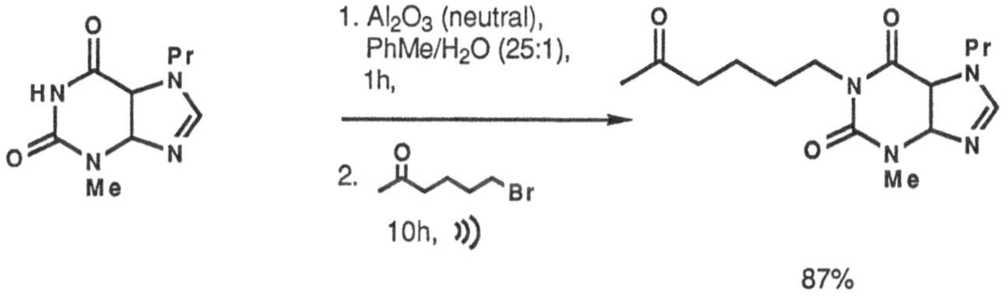

1. Al$_2$O$_3$ (neutral),
   PhMe/H$_2$O (25:1),
   1h,

2. [structure] Br

10h,)))

87%

Scheme 82

N-alkylation of *N,N*-diazacoronands can also be carried out under exceptionally mild conditions using aqueous KOH as the base (Scheme 83). In contrast, the literature methods for methylation of such species normally require drastic conditions which are not compatible with highly functionalized substrates, such as those which bear optically active substituents.

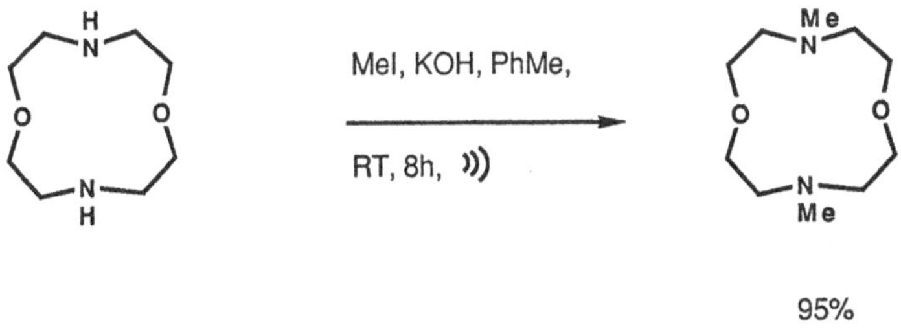

MeI, KOH, PhMe,

RT, 8h,)))

95%

Scheme 83

Initial investigations showed that use of a phase transfer catalyst was unnecessary. However, it later transpired that the reaction was auto-catalysed although addition of an independent catalyst was necessary in some cases [182].

Long chain α-bromo-ω-carboxylic acids were converted to their 128-iodo-derivatives in quantitative yield simply by sonolysis of the acid with 128-iodine in ethyl acetate solution for 20 min [183]. Furthermore, investigations showed that the system would tolerate up to 7.5% volume water. As yet there are no examples of this simple procedure being used in place of the commonly used Finkelstein method [184] for preparation of alkyl iodides from the equivalent bromide.

Introduction of deuterium or tritium into aromatic compounds can also be carried out with a surprising degree of selectivity that is not seen in the equivalent thermal reaction. Thus reaction of p-bromoacetylbenzene with NaOD—D$_2$O in the presence of Raney nickel gives an 83% yield of the product resulting from direct displacement of the halide on exposure to ultrasound [185]. This is in stark contrast to the thermal reaction which gives a mixtures of products in which the trideuterated arene predominates (Scheme 84).

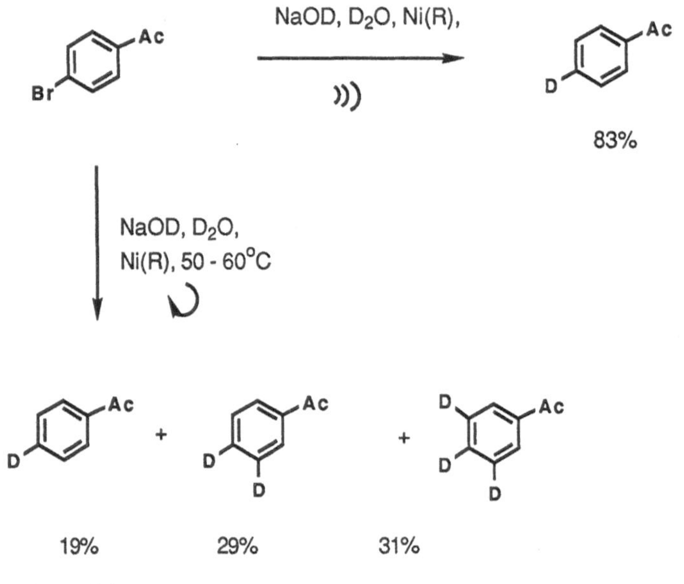

Scheme 84

^{13}C-labelled aminolevulinic acid containing >90% ^{13}C at C5 was prepared by treatment of 3-carboethoxyproponyl with Cu^{13}CN. The intermediate acyl-nitrile was reduced by treatment with zinc in AcOH/Ac$_2$O. Sonolysis of the reaction mixture gave the N-acetylated ester in 98% yield [186].

A further example of the use of ultrasound to simplify a standard organic procedure is the discovery that a variety of sugar acetals can be prepared by the acid catalysed reaction of the sugar with acetone in a fraction of the time necessary for the equivalent thermal reaction. In addition, yields of the diacetals were typically 20–30% higher [187] (Scheme 85). The authors also report that neutralisation of the acidic reaction mixture with solid potassium or sodium carbonate was very fast, probably due to more efficient dispersion of the reactants. However, this process was executed in a cleaning bath and was, not surprisingly, ineffective when working on more than 0.2 moles of material. A noted limitation was that no rate increase was observed in the case of D-ribose.

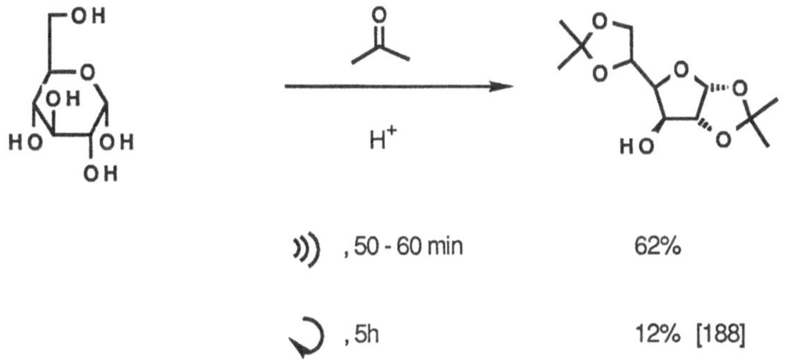

$$)) \quad , 50 - 60 \text{ min} \qquad 62\%$$

$$\circlearrowleft \quad , 5h \qquad 12\% \text{ [188]}$$

Scheme 85

Synthesis of spirocyclic ketones can be achieved by the thermal reaction of cyclic ketones with α,ω-dibromoalkanes in the presence of a base [189, 190]. However, these conditions simply promote rapid self-condensation in the case of cyclopentanone [189] restricting the general applicability of the method. In contrast, sonolysis of the reaction mixture gives good yields of spiro-(4,*n*)-alkan-1-ones [191] (Scheme 86).

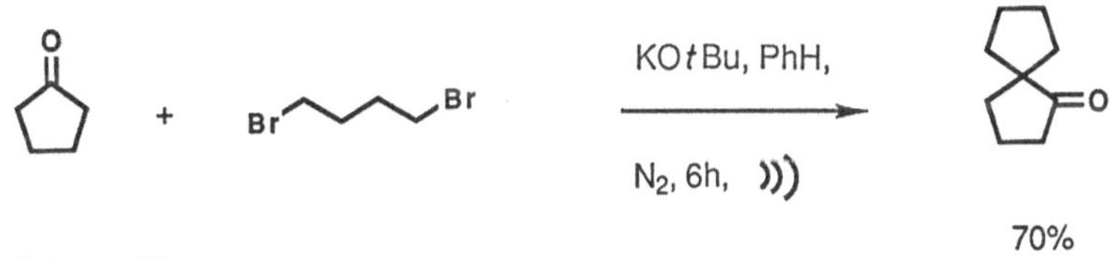

Scheme 86

Synthesis of a type I cyanolipid was achieved via the intermediate (**42**) Reaction of the allylic bromide (**41**) with the sodium salt of the required long-chain fatty acid occurred smoothly in the presence of a catalytic amount of tetrabutylammonium bromide [192] (Scheme 87).

Preparation of thioamides from amides was also shown to proceed faster and more efficiently under ultrasonic conditions (Scheme 88). Yields of between 77 and 97% were obtained within 1–2 h [193]. In contrast, the standard reaction involves use of a large excess of reagent as a result of the insolubility of P_4S_{10}, and results obtained using Lawesson's reagent tend to be unpredictable.

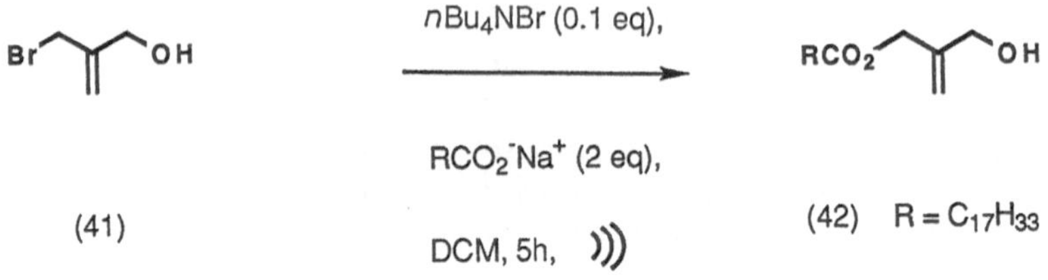

Br⌁CH₂⌁OH

nBu_4NBr (0.1 eq),

$RCO_2^-Na^+$ (2 eq),

DCM, 5h,)))

RCO₂⌁OH

(41)

(42)   R = $C_{17}H_{33}$

Scheme 87

P_4S_{10} (1-1.5 eq),

RCONH₂

1-2h, THF,)))

30 - 40°C

RCSNH₂

77 - 97%

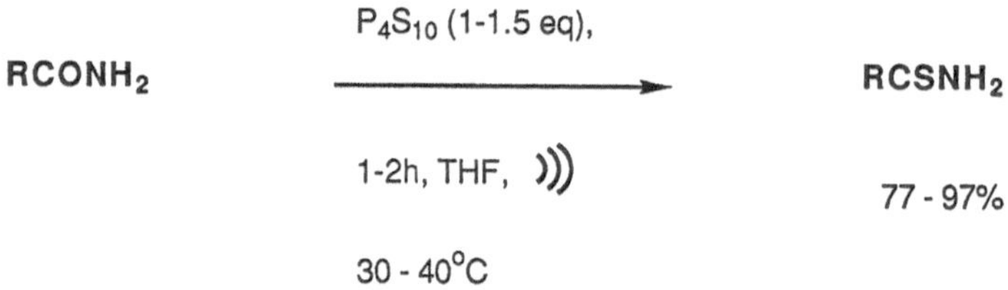

Scheme 88

Ultrasound can also be used to promote formation of alkylthiocyanates [194] and thiocarbamates [195] by direct reaction of alkyl halides with the relevent sulphur containing anion (Scheme 89).

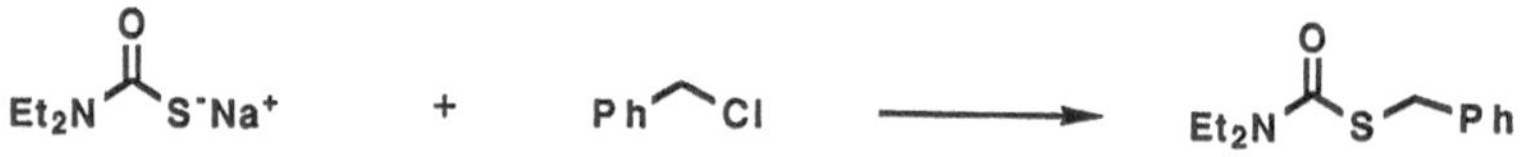

Et₂N—C(=O)—S⁻Na⁺   +   Ph⌁Cl   ⟶   Et₂N—C(=O)—S⌁Ph

Scheme 89

O-alkyl-3-oxothionoesters were prepared by reaction of the sodium enolate with $(EtOCS)_2S$ in a ratio of 2.1:1. Hydrolysis of the intermediate gives the oxothionoester. However, the products are often contaminated with the products of ketone self condensation. Conversely, sonication of the reaction mixture gave good yields of the pure product in this case [196] (Scheme 90).

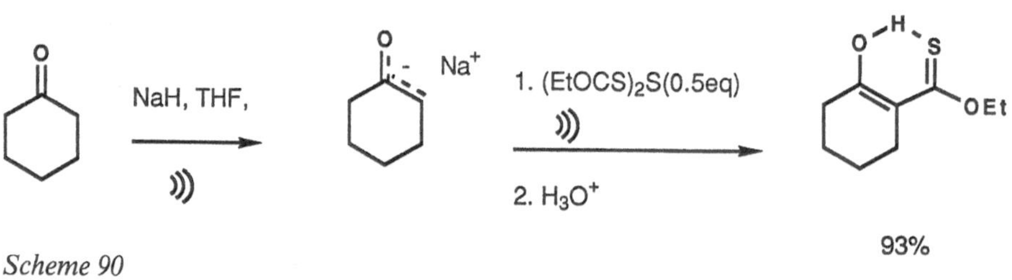

NaH, THF,

)))

Na⁺

1. $(EtOCS)_2S$ (0.5eq)

)))

2. H_3O^+

93%

Scheme 90

Ferrocene-containing pyrimidine derivatives have been prepared by addition of thiourea to arylferrocenylpropenones. Heating the reaction components at 60 to 70 °C for 3–5 h gave complex mixtures of products. However, sonication gives the product shown in good yield [197] (Scheme 91).

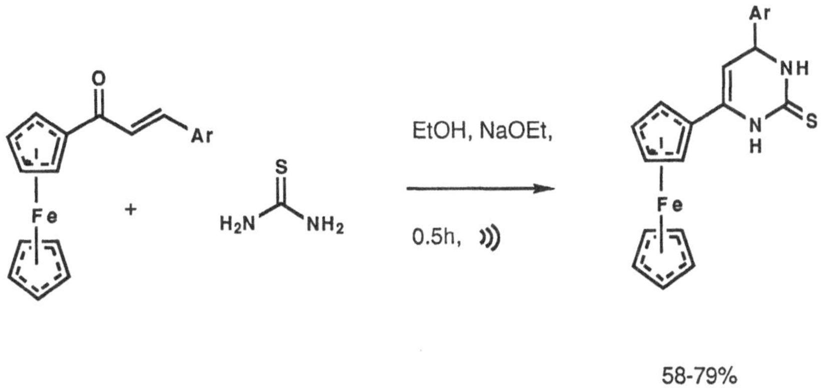

Scheme 91

A Japanese patent describes the preparation of *N*-formyl aspartic anhydride by direct reaction of L-aspartic acid with formic acid in the presence of acetic anhydride and an optional catalyst. Reaction is complete after 12 h of sonolysis at 45 °C and gives a 93% isolated yield of the anhydride [198].

Cyclic carbonates are available by reaction of terminal epoxides with CO_2 in the presence of triethylamine. The reaction is catalysed by both organotin and antimony halides. Reaction for 7 h at 35 °C gives a 46% yield of the carbonate (**43**). By contrast, exposing the reaction mixture to ultrasound increases this to 79% over the same time period [199] (Scheme 92).

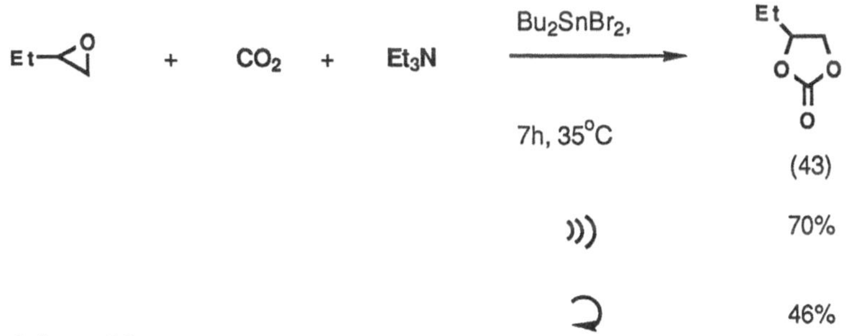

Scheme 92

Previous sections have detailed numerous examples where ultrasound has been used to activate a metal surface. Reaction with species in solution then provides a rapid route to a variety of organometallic products. However, these effects are not confined to organometallic systems and there are already a number of examples where inorganic bases can be used under heterogeneous conditions in wholly organic reactions. Shibata and-co-workers [200] have described the cyanomethylation of a variety of chalcones by Michael addition of the anion derived from acetonitrile. Sonolysis for 15 min in the presence of KO_2 gave the adducts (**44**) as the major product in each case (Scheme 93).

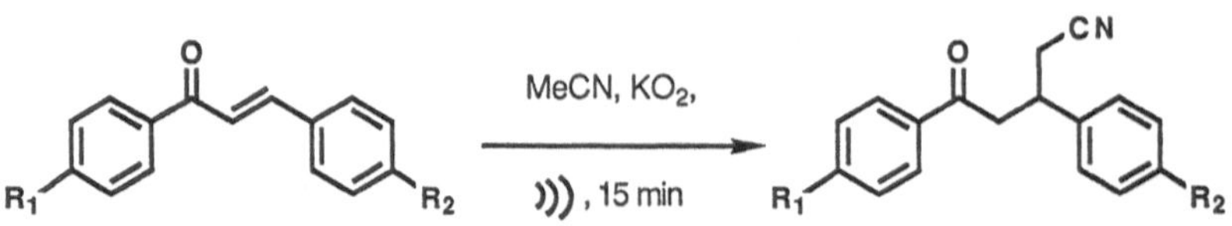

R$_1$, R$_2$ = H, OMe, Me, Cl

(44) 33-67%

Scheme 93

Ultrasound can also be used to promote the Cannizarro reaction between aldehydes and a variety of solid bases in 96% ethanol. These included commercial barium hydroxide C—O (Ba(OH)$_2$ · 8 H$_2$O, Probus SA) and C-200, an activated barium hydroxide catalyst [201]. The latter is known to catalyse a number of other organic reactions, e.g. Michael addition, the Claisen-Schmidt reaction and aldol condensations, but has no effect on the Cannizzaro reaction under normal thermochemical conditions. Nevertheless, exposing the reaction mixture to ultrasound for 10 min produces quantitative yields of the respective products in most cases. No reaction is observed in the absence of the catalyst, demonstrating that formation of radicals by interaction with the solvent has no effect on the ongoing reaction. However, the reaction was also shown to be independant of the polarity of the solvent, in contrast to other reactions catalysed by the basic centres of C-200. Further investigation showed that poisoning the reductive sites of the catalyst with 1,3-dinitrobenzene completely inhibited its action. This led the authors to propose that reaction occurs by single electron transfer from the reductive sites on the surface of the catalyst. However, this failed to account for the 'catalytic' nature of the process as it implied that the reductive sites must be transformed into basic sites during the course of the reaction. This mechanism was later revised when further investigations showed that the reaction was truly catalytic with respect to the reductive sites of the C-200 [202]. This suggests that ultrasound plays a role in the initial SET, or it may simply behave as an initiator. The effects of ultrasound on increasing the number of initiation sites and/or aiding desorption from the surface do not seem to have been considered.

C-200 can also be used to catalyse the Wittig-Horner reaction between aldehydes and diethylphosphonoacetate [203] (Scheme 94). Sonication of the reaction mixture allows the quantity of catalyst used to be reduced. Reaction takes between 5 and 30 minutes and is carried out at ambient temperature, in contrast to the thermal reaction which is carried out at 70 °C. The presence of a small amount of water has been shown to be beneficial to both the thermal and ultrasonic reactions. However, its role is disputed and two theories have been put forward to account for this. The first of these suggests that water destroys the 1,2-oxaphosphetane intermediate in a similar way to that observed in the presence of crown ethers [204] or that it stabilises the Ba(OH)$_2$ · H$_2$O structure of the C-200. The authors favour the latter argument on the grounds that this has already been established for the thermochemical Wittig-Horner reaction [205].

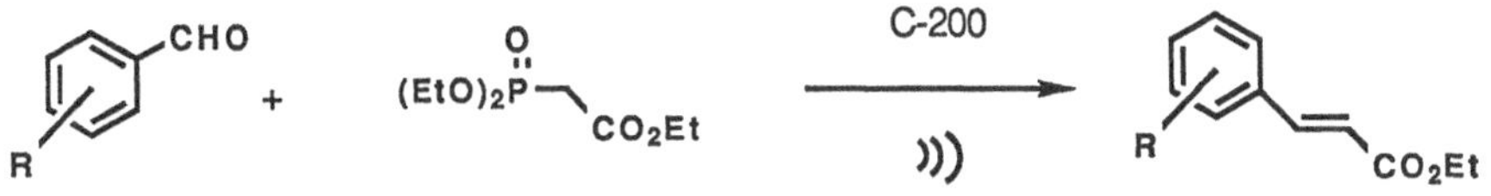

Scheme 94

The stereochemical outcome of the reaction is also identical to that of the thermal reaction.

A short investigation in our laboratories showed that ultrasound could be used to vastly improve the yields of conjugated dienes obtained using the standard Wittig reaction [206] (Scheme 95). Deprotonation of the phosphonium salt (**45**) by *n*-butyllithium occurs extremely slowly, and quenching with benzaldehyde gave a meagre 8% of the diene. The yield obtained using LDA was considerably higher (79%). However, close observation suggested that the insolubility of the phosphonium salt in THF was a major barrier to the formation of the Wittig reagent. Hence, we repeated the initial experiment in the presence ultrasound and found that formation of the deep red phosphorous ylide was complete after 1 h. Quenching the reaction mixture with benzaldehyde then gave a remarkable 91% yield of the *cis* and *trans* dienes in a ratio of 2:3. This represents a substantial improvement on any of the available literature methods, relieves us of the need to preform LDA in situ, and produces a degree of selectivity in favour of the E,E-diene. Hence, this appears to be the method of choice for isoprenylation of aldehydes via Wittig methodology, although its general applicability to the synthesis of conjugated dienes remains untested.

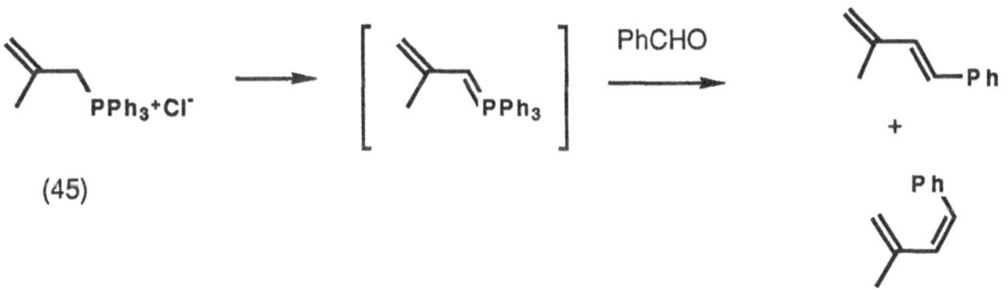

(45)

Deprotonation conditions	Yield of diene	Ratio of *cis* : *trans* diene
1. *n*BuLi, RT, 20h	8%	1 : 1
2. LDA, RT, 2h	79%	1 : 2.3
3. *n*BuLi, 1h,)))	91%	2 : 3

Scheme 95

Sonolysis of powdered sodium hydroxide and chloroform with concomitant stirring leads to generation of dichlorocarbene. This could be trapped by a variety of olefins providing a high-yielding synthesis of dichlorocyclopropanes [34] (Scheme 96). Competition experiments showed that the carbene produced was selective for highly substituted double bonds and that its behaviour was analogous to that produced in potassium t-butoxide/chloroform systems. Interestingly, sonication or stirring alone do not produce such a marked effect. This is probably due to the increase in mass transport effected by the stirring. The reaction is carried out in a low intensity cleaning bath and the authors had already remarked that large scale reactions (i.e. >5 mmol of alkene) gave poor yields of cyclopropane on this account. However, its major advantage is that the reaction proceeds without the need for the phase transfer catalysts that are normally required and this modification considerably simplifies the work-up procedure.

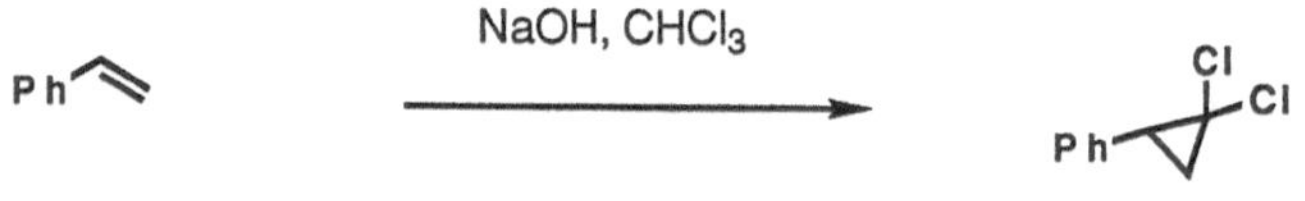

Conditions used		Reaction Time/h	% Yield
1.	)))	1.5	95
2.	—	16	31
3. —	)))	20	38

Scheme 96

A number of liquid/liquid biphasic systems have also been shown to benefit from ultrasonic emulsification. These include the formation of organic azides from activated primary chlorides and bromides [207] and saponification of aryl esters [5] and nitriles [208] to give the corresponding carboxylic acids.

Base-catalysed formation of ethers and esters has been carried out by reaction of the relevant alcohol or carboxylate with an alkyl halide in the presence of a phase transfer catalyst such as PEG methyl ether or nBu_4Cl. The results of these investigations are summarised below [209] (Scheme 97). The esterification reaction is particularly suited to reaction with highly lipophilic acids, when the products can be obtained in high yield, uncontaminated by side products.

A new method for the glycosidation of 2-benzenesulphonyl tetrahydropyrans has been developed. In most examples, reaction was complete after

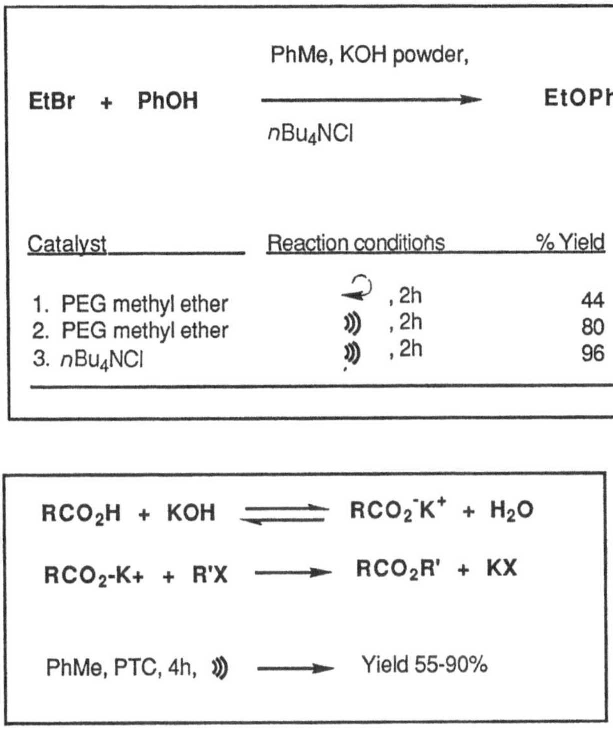

EtBr + PhOH $\xrightarrow{\text{PhMe, KOH powder,}}$ EtOPh

nBu$_4$NCl

Catalyst	Reaction conditions	% Yield
1. PEG methyl ether	↻ , 2h	44
2. PEG methyl ether	))) , 2h	80
3. nBu$_4$NCl	))) , 2h	96

$$RCO_2H + KOH \rightleftharpoons RCO_2^-K^+ + H_2O$$

$$RCO_2\text{-}K+ + R'X \longrightarrow RCO_2R' + KX$$

PhMe, PTC, 4h,))) $\longrightarrow$ Yield 55-90%

Scheme 97

15–24 h at RT in the presence of MgBr$_2 \cdot$ Et$_2$O and NaHCO$_3$. However, exposing the reaction mixture to ultrasound produced a dramatic rate enhancement in cases where the reaction was sluggish and high yields could be obtained under particularly mild conditions [210] (Scheme 98).

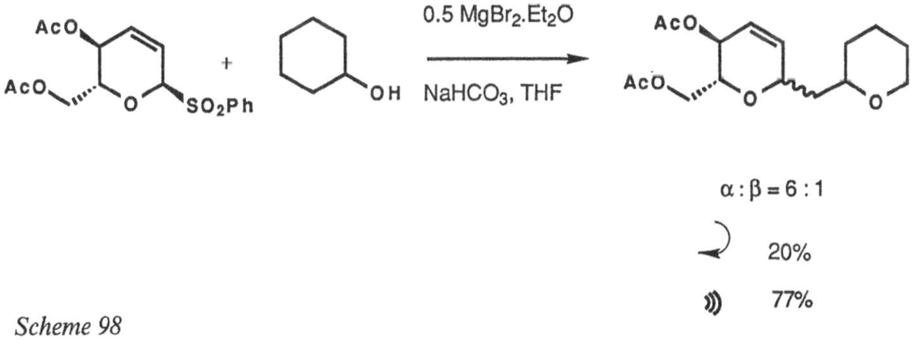

$$\alpha : \beta = 6 : 1$$

↻   20%

)))   77%

Scheme 98

Application of ultrasound to a number of alumina catalysed aldol reactions has also been found to be beneficial. Initial reports stated that self-condensation of ketones was enhanced by a factor of two over the acid catalysed dimerisation [211].

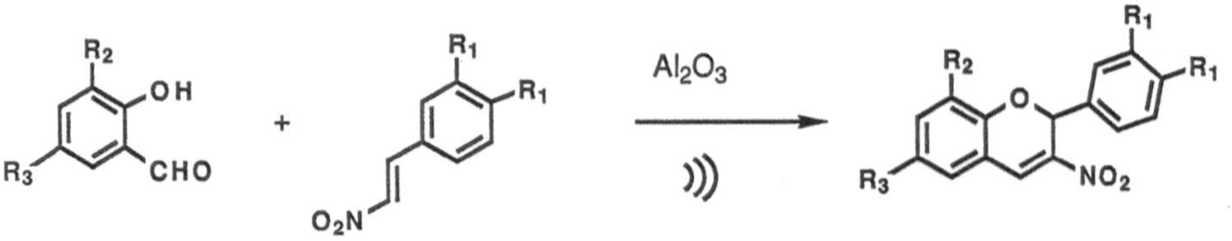

Scheme 99

This was later exploited in a synthesis of 3-nitro-2H-chromenes [212] with a variety of electron withdrawing substituents in moderate to good yield (Scheme 99). In 1983 Ando et al. [117] had reported the synthesis of aroyl nitriles from the corresponding chlorides by direct displacement with potassium cyanide [213].

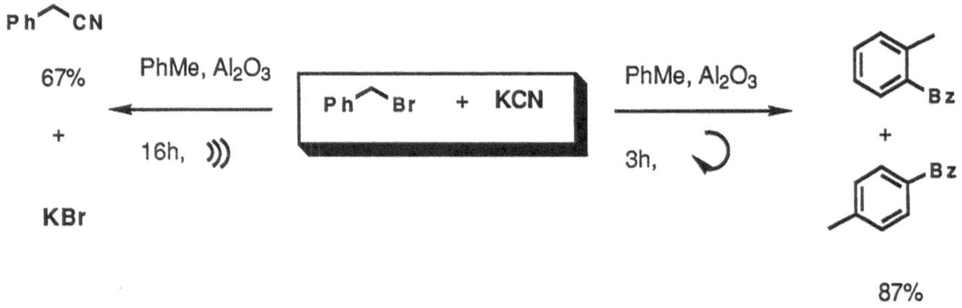

Scheme 100

Extending this to the synthesis of aryl nitriles they found that yields obtained were low (27%) but could be significantly augmented by the presence of a basic alumina catalyst and a trace of water [214, 215] (Scheme 100). However, the stirred reaction gives a mixture of *o*- and *p*-benzyltoluene resulting from Friedel-Crafts attack on the solvent. This variation in reaction pathway appears to be due to the formation of active ion aggregates on sonolysis of the alumina which block those surface sites normally associated with its Lewis acid nature, coupled with the enhanced nucleophilicity of the cyanide anion. Further examination of this effect [216] showed that ultrasound was essential for nucleophilic substitution to take place. Both the Friedel-Crafts reaction and the cyanide substitution were retarded when sonication was ceased after pre-treatment of the KCN/alumina mixture was complete, but before addition of the bromide. Ultrasonic irradiation was used in place of a phase transfer catalyst to effect simultaneous deacyloxylation and dehydrogenation of terphthaloyl nitriles. The intermediates were subsequently treated with bromine and triphenylphosphine to give moderate yields of the substituted TCNQ derivatives [217] (Scheme 101).

Brown and Racherla have also exploited ultrasound to speed up hydroboration of substrates which react extremely slowly with a wide variety of hydroborating agents [76a]. Of particular note is the reaction of (—)-α-pinene

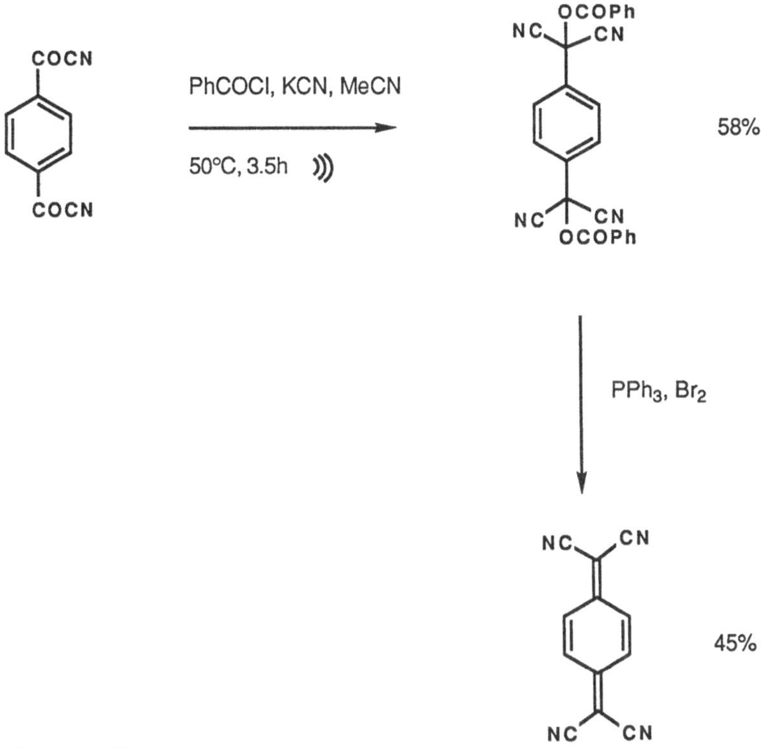

Scheme 101

with 9-BBN to give alpineborane (**46**) — a useful reagent for the asymmetric reduction of α,β-acetylenic ketones and other prochiral carbonyl compounds (Scheme 102). The authors report that the reaction takes 12 h in THF at 65 °C, although this can be reduced to 1 h by using the neat reagents which gives a quantitative yield of alpine-borane. Once again, this is in complete accordance with the predictions of the "hot spot" theory.

Scheme 102

   In situ generation of Grignard reagents also allows access to a variety of trialkylorganoboranes which are unavailable by hydroboration. For example triisopropyl, benzyl, phenyl and allylboranes can be prepared in 10–30 min by sonolysis of a mixture of boron trifluoride diethyletherate, magnesium turnings and a crystal of iodine in ether. Dropwise addition of the alkyl halide gives the desired product within minutes without the need for pre-formation of the Grignard reagent [76b] (Scheme 103).

$$3nBuBr \xrightarrow[\text{Et}_2\text{O, 15 min}]{3Mg\ BF_3.OEt_2\ \text{)))}} nBu_3B$$

Scheme 103

Casiraghi and co-workers have reported a highly stereocontrolled synthesis of both *syn* and *anti*-1-(2-hydroxyaryl)glycerol derivatives by reaction of Mg^{2+} and Ti^{4+}-phenolates, generated in situ, with the appropriate enantiomer of 2,3-*O*-isopropylideneglyceraldehyde under ultrasonic irradiation [218]. In all cases the products were isolated with a diastereoisomeric excess of greater than 90% (Scheme 104). Initial investigations were carried out using $MgBr^+$ as the counterion in ethereal solution and gave the products of *syn* addition in high yield. In contrast, the stirred reaction gave a greatly reduced yield of the addition product with a lower degree of stereoselectivity (85:15). The stereochemistry of the addition was completely reversed by using $Ti(OiPr)_3^+$ as the counterion in toluene. This gave the corresponding *anti*-glycerols with similarly high diastereoselectivity and in good yield. Hence, this provides a promising route to new chiral alditols bearing ring-hydroxylated appendages.

α-Cyanobenzoylarylsulphonates were prepared in a two-phase water/toluene (1:2) system (Scheme 105). Optimal conditions involved sonication at 28 kHz and 600 W every 20 minutes over a 5 h period. Use of a phase transfer catalyst was unnecessary under these conditions and gave a product "of greater purity" than that which could be obtained under thermal conditions in the nBu_4NBr catalysed reaction [219].

Olah and co-workers have reported a rapid and high-yielding synthesis of adamantanes by sonication of the corresponding polycyclic hydrocarbon precursors in the presence of a superacid, CF_3SO_3H—SbF_5. Thus 98% yields of both adamantane and diadamantane could be obtained within 1.75 h. The corresponding thermal rearrangement took 10 h at ambient temperature and the overall yield of adamantane was slightly lower (74%). However, sonication reduced the reaction time from 98 to 8 h [220] (Scheme 106).

Moon and co-workers have made a detailed study of the two-phase basic hydrolysis of aromatic carboxylic acid esters [5] and the corresponding acid-catalysed reaction has been described for the hydrolysis of sugar cane bagasse. Optimal conditions involved the use of 38% concentrated hydrochloric acid in combination with lithium chloride [221], which produced a 62% rate increase in comparison with the stirred reaction.

Examples of homogeneous reactions which can be accelerated using a basic cleaning bath are few and far between. However, it has been reported that aminocyanation of hexahydrobenzo[a]quinolizinone can be carried out in acetic acid solution on irradiation with ultrasound [222] (Scheme 107). In contrast, the thermal reaction took between 12 and 13 days and isolated yields were appreciably lower. These reaction conditions were developed when reaction under the usual Strecker conditions, that is reaction with the

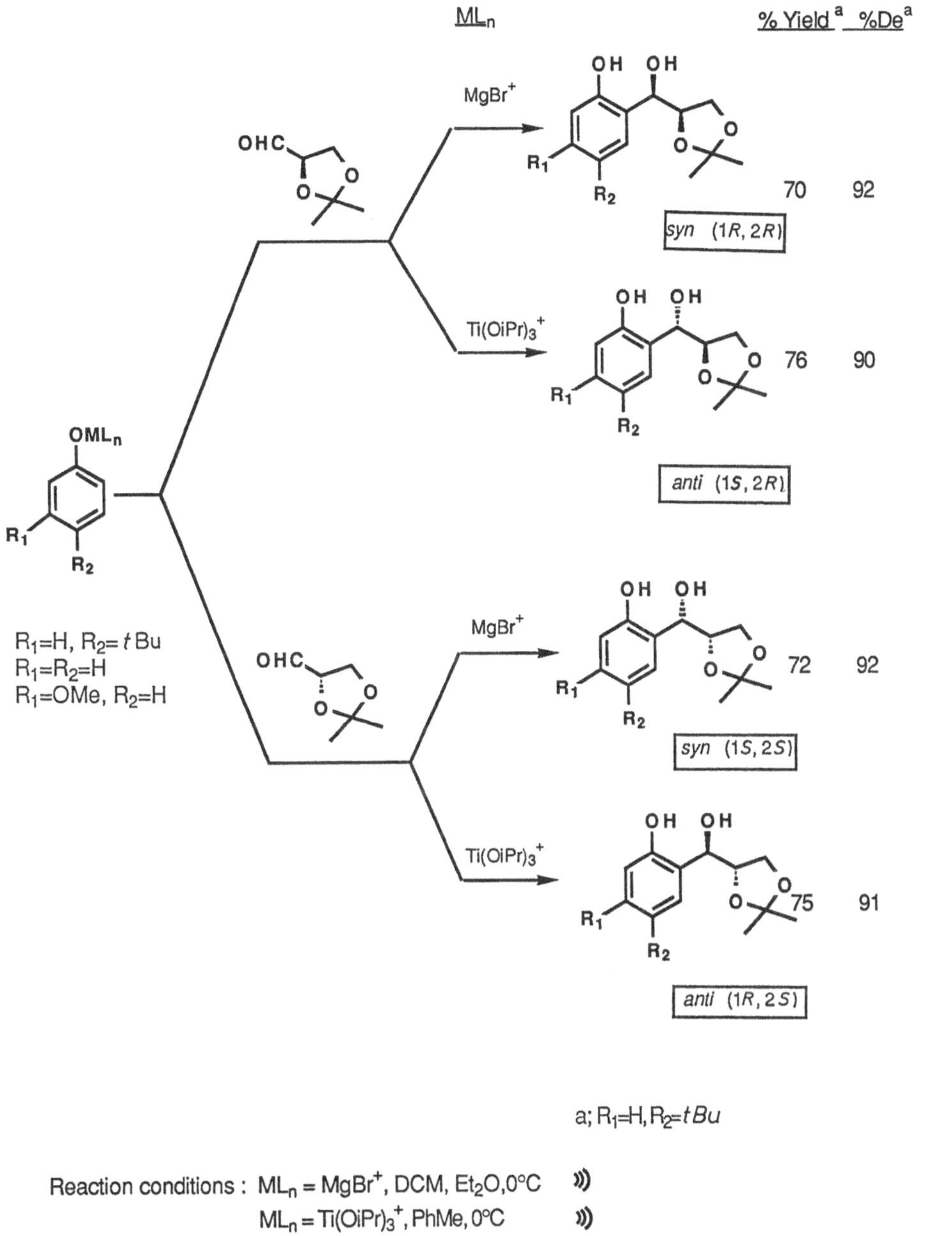

Scheme 104

relevant amine and potassium cyanide in aqueous ethanol, gave the undesired cyanohydrin.

Dehydration of the *N*-nitrosocompound (**47**) in acetic acid gives the corresponding sydnone (**48**) (Scheme 108). The cyclisation process nor-

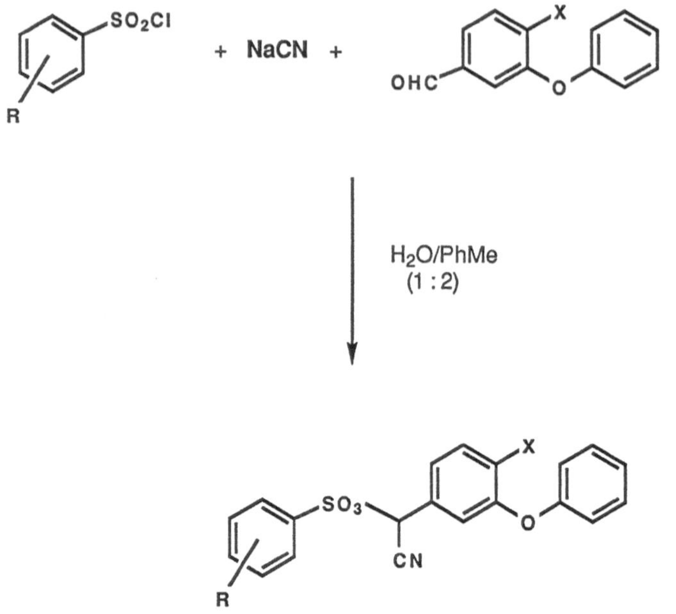

Reaction conditions	% Yield	% Purity
5h,)))	99	94
nBu$_4$NBr, ⊃	92	45

Scheme 105

mally requires several days to reach completion. In contrast, sonication reduces this to a few hours [223]. Transformation of the monosydnone (**48**) to the disydnone (**49**) can also be effected in trifluoroacetic anhydride [224]. The reaction is equally fast, although the overall yield is somewhat lower.

Finally, ultrasound has been used in the preparation of a variety of organoselenium and organotellurium compounds by reaction of alkyl halides, tosylates, mesylates and epoxides with the phenylselenium anion [107] and direct reaction with the Se_2^{2-}, Se^{2-}, Te_2^{2-} and Te^{2-} anions produced by electrochemical reduction of the insoluble elements [35], or by treatment with an alkali metal in the presence of 10 mol% naphthalene [225]. This effectively "solvates" the metal by complexation in an analogous manner to that observed with magnesium [66]. Sonication of the reaction mixture halves the time with respect to the stirred reaction at ambient temperature. However, this is one of the few literature examples where the reaction is actually faster in refluxing THF than under the influence of ultrasound (Scheme 109).

In each case the metal used was in the form of small chips and it seems likely that the limited power of the cleaning bath contributed to the observed

	Reaction time/h	% Yield adamantanoids		
		ada	*diada*	*triada*
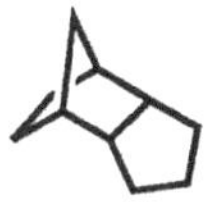	1.5	98		
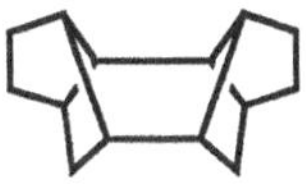	1.75		98	
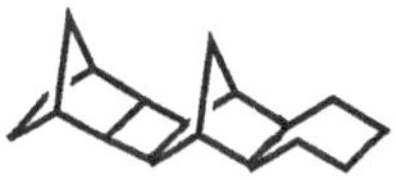 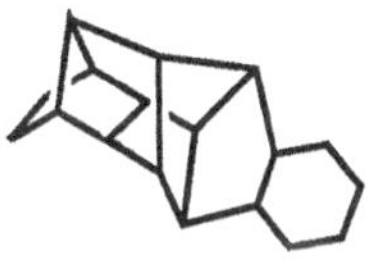	8.0			74

Scheme 106

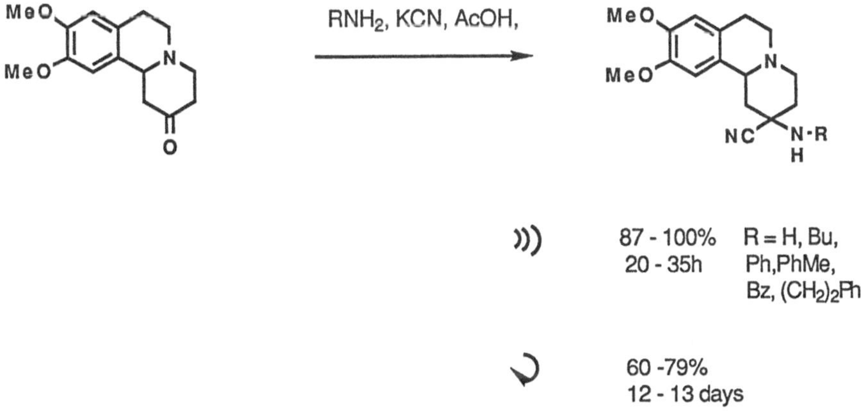

RNH₂, KCN, AcOH,

))   87 - 100%   R = H, Bu,
20 - 35h   Ph,PhMe,
Bz, (CH₂)₂Ph

↺   60 -79%
12 - 13 days

Scheme 107

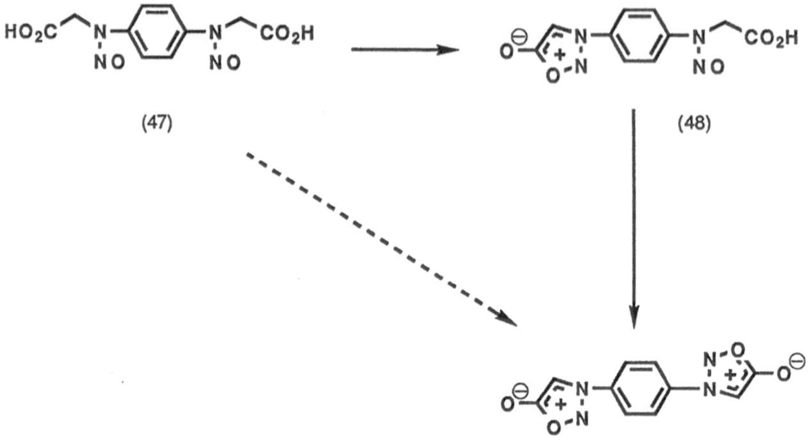

(47)

(48)

(49)

Transformation	Reaction conditions	% Yield
(47) to (48)	Ac₂O, RT, 10 days ⤳ Ac₂O, 5h ⑅	"very low yield" 98
(48) to (49)	Ac₂O, RT, 5days, ⤳ Ac₂O, 2h, ⑅ TFAA, 2h, RT ⤳	17 90 71
(47) to (49)	Ac₂O, RT, 5 days, ⤳ Ac₂O, 7h, ⑅	17 98

Scheme 108

$$2M \quad + \quad 2Se \xrightarrow{\substack{THF, \\ \\ naphthalene \\ (10\ mol\%)}} M_2Se_2$$

M	reflux	ultrasound/0-5°C	RT/stir
K	40 min	85 min	150 min
Na	65 min	65 min	135 min
Li	8h	9.5h	15h

Scheme 109

94

result. It is possible that use of the commercially available dispersions of alkali metals would change the overall outcome.

Sodium phenylselenide was prepared in 1 h by sonolysis of diphenyldiselenide with sodium in the presence of benzopheneone, which acts as an electron transfer agent (Scheme 110) [107]. The reaction between diphenyldiselenide and solid sodium is virtually negligible at room temperature. However, initial studies showed that the reaction could be brought to completion within four days in the presence of ultrasound. A brief investigation of the effect of solvent on the reaction was carried out in line with those described by Luche and co-workers [83]. Thus it was discovered that the reaction time could be halved by using xylene in place of THF. However, from a practical point of view, the difference in boiling points between that of xylene and THF is considerable. This would severely restrict the applicability of the method as isolation of volatile or thermally unstable selenides would be virtually impossible.

The greatest enhancement in the rate of reaction was observed when a catalytic amount of benzophenone was added to the reaction mixture. This modification had the added advantage that the reaction became self-indicating as a result of the characteristically dark blue sodium benzophenone ketyl radical. Dropwise addition of a solution of diphenyldiselenide in a minimal volume of THF appeared to result in virtually instantaneous formation of the cream-coloured sodium phenylselenide and the reaction was complete when the last mauve tinges had disappeared.

This obviates the need to use complicated reagents and the odiferous, unpleasant reagent is prepared in a sealed flask ready for use and can be transferred directly to the reaction vessel via a syringe or cannula. Sodium methane sulphonate is equally available under analogous conditions [226].

Yields of phenylselenol obtained using this reagent were virtually quantitative — even in cases where competing reactions might have been expected to prove problematic. For example, a fragment of the ionophore antibiotic M139603 (**50**) was prepared using this methodology [227] (Scheme 111).

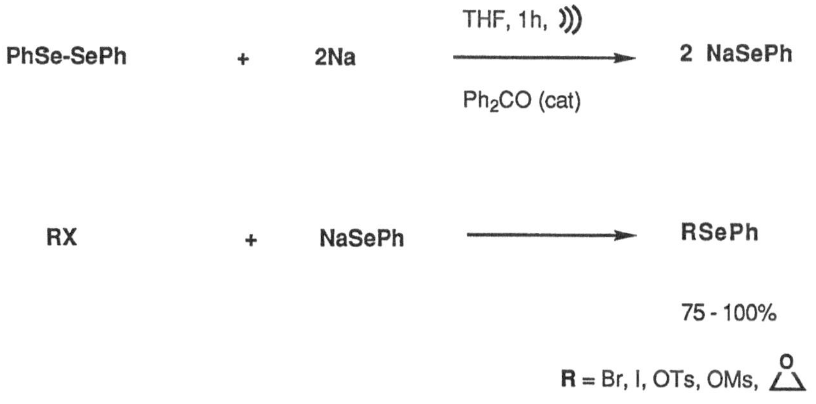

Scheme 110

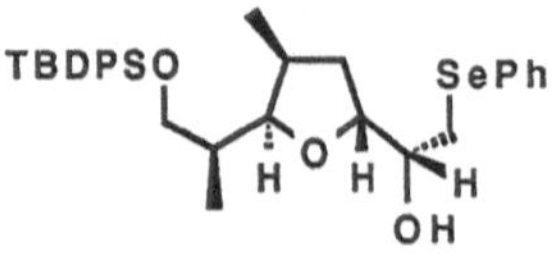

(50)

Scheme 111

Furthermore, treatment of ultrasonically generated sodium phenylselenide with acid gives the parent selenol which, although commercially available, is currently more than six times the price of diphenyldiselenide.

Sequential synthesis of the E_2^{2-} and E^{2-} anions by electrochemical reduction of the elements Se or Te provides a route to both dialkyl-di- and monoselenides and tellurides [35]. Ultrasound has also been used to promote the electrochemical synthesis of unsymmetrical diaryl chalcogenides typified by 4-phenylseleno- and -tellurobenzonitrile [228]. However, the rate enhancements observed when electrochemical reactions are exposed to ultrasound would appear to stem from the increase in mass transport rather than cavitational effects [229].

12 Ultrasonic Acceleration of Redox Reactions

The well-documented effects of ultrasound on heterogeneous systems have been exploited to produce beneficial increases in the rates of a number of reactions that would more commonly be executed under homogeneous conditions. For instance, the reduction of aryl halides with lithium aluminium hydride occurs extremely slowly in THF solution. However, Han and Boudjouk have reported that the reaction can be carried out in a fraction of the time if the reaction mixture is sonicated in dimethoxyethane (DME) solution [6]. Sonolysis of the heterogeneous reaction mixture gives high yields (70–99%) of the reduced product. The effect was particularly marked in the case of deactivated aryl halides; for example, p-bromotoluene gave a 97% yield of toluene in 5 h. In contrast, the yield of the stirred reaction was a modest 21% after 24 h at room temperature in THF (Scheme 112).

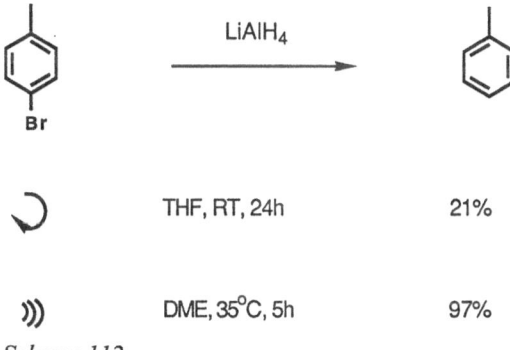

Scheme 112

This discovery was exploited by Lukevics et al. [7], providing a route to the hydrides of a number of group IV element derivatives from the corresponding halo, alkoxy and amino compounds. The reaction is completely dependent on sonication and does not occur in its absence (Scheme 113).

Sonolysis of a variety of alkyl substituted nitrobenzenes under an oxygen atmosphere in the presence of solid potassium hydroxide and a polyethylene glycol phase transfer catalyst gave moderate yields of the corresponding carboxylic acid salt (51). Interestingly, the stirred reaction gave no acid and the products observed, (52) and (53), arose from dimerisation of the starting material [230] (Scheme 114). This is one of the few cases where the presence of ultrasound actually changes the pathway of a reaction.

97

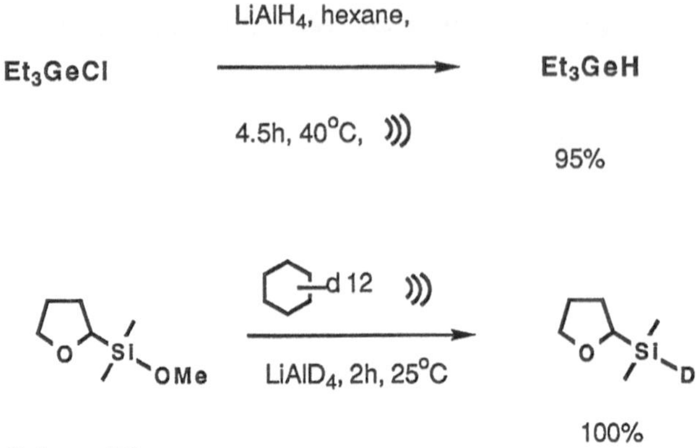
LiAlH4, hexane,
Et3GeCl
Et3GeH
4.5h, 40°C,
95%
d 12
OMe
LiAlD4, 2h, 25°C
D
100%

Scheme 113

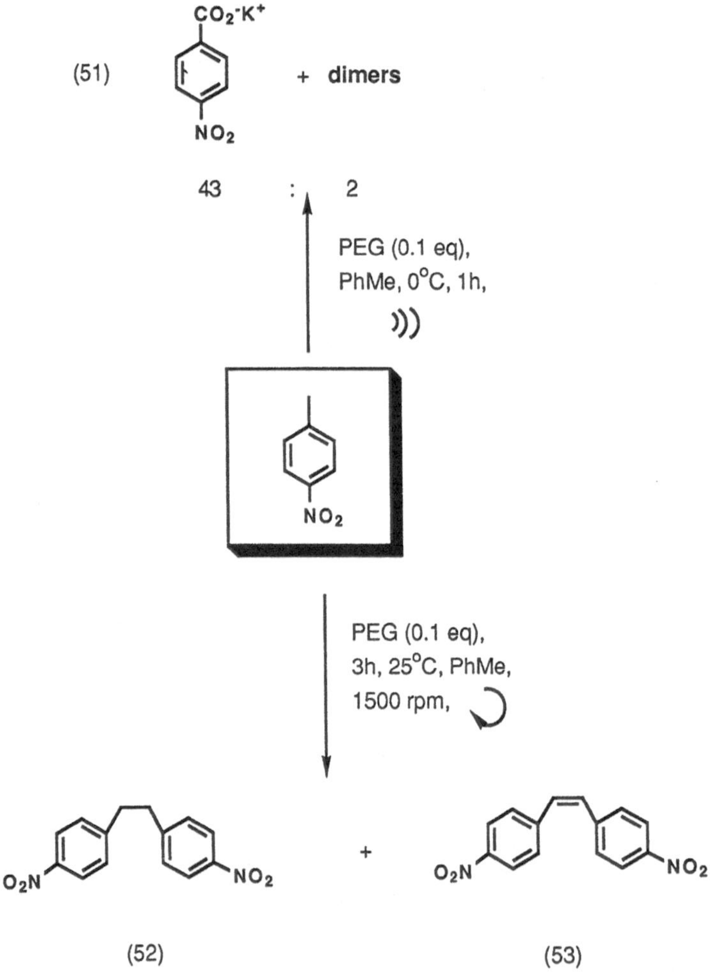
(51)
CO2-K+
+ dimers
NO2
43 : 2
PEG (0.1 eq),
PhMe, 0°C, 1h,
NO2
PEG (0.1 eq),
3h, 25°C, PhMe,
1500 rpm,
O2N
NO2
+
O2N
NO2
(52)
(53)

Scheme 114

Studies on the homogeneous system had previously shown that the acid was the predominant product if the concentration of the reagents was low, or the partial pressure of oxygen was high [231]. These conditions are satisfied if one assumes that the involatility of potassium hydroxide leads to an extremely low concentration of base in the vapour phase, that is, at the site at which reaction occurs.

Oxidation of secondary alcohols to ketones can also be effected in good yield using solid potassium permanganate [232] in hydrocarbon solvents (Scheme 115). However, reaction with aryl compounds was slow and primary alcohols gave mixtures of the aldehyde and the corresponding carboxylic acid. The authors also noted that, whilst the stirred reaction was strongly dependent on the polarity of the solvent, reaction in the presence of ultrasound showed very little variation. The slight drop in yield observed is presumably due to the increasing volatility of the solvent used rather than its polarity.

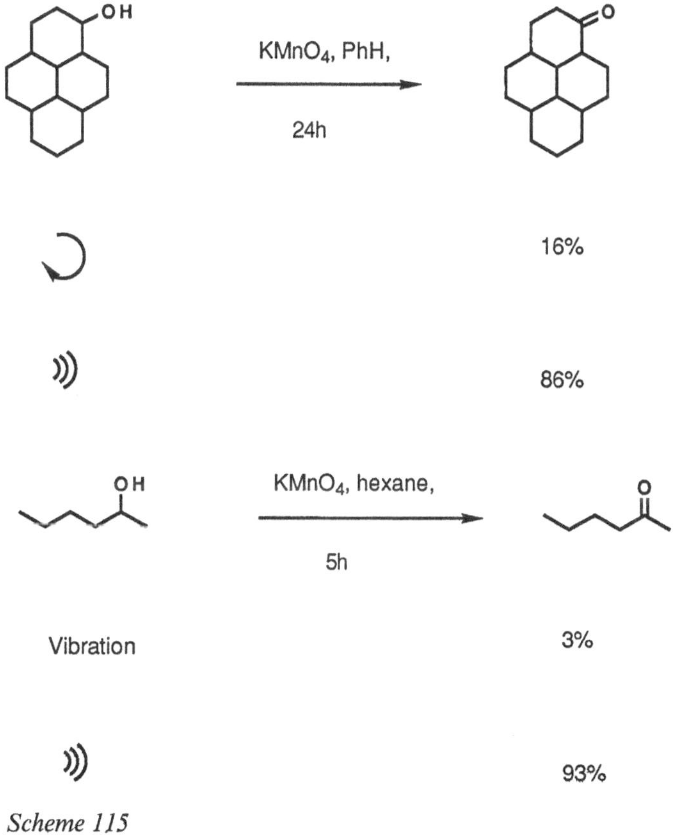

Scheme 115

Similar results were later obtained by Han and co-workers using $BaMnO_4$ and $KMnO_4 \cdot CuSO_4 \cdot 5 H_2O$ [233].

Aromatic nitro-groups can be reduced to give the equivalent aniline at room temperature and atmospheric pressure using iron, activated carbon, and hydrazine hydrate as the hydrogen source. High yields of aniline were obtained under these mild conditions (Scheme 116).

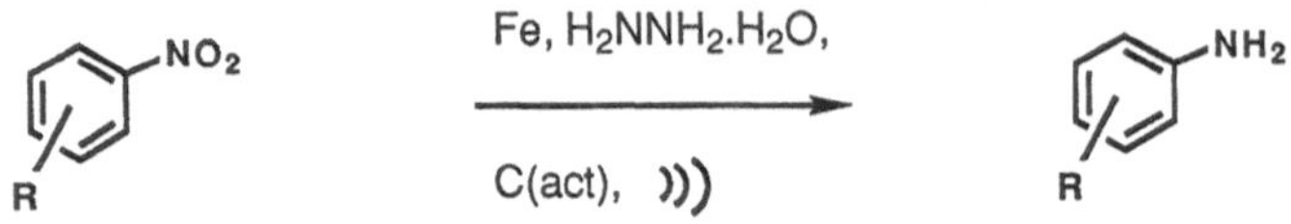

Fe, $H_2NNH_2 \cdot H_2O$,

C(act),)))

87 - 95%

Scheme 116

The conversion yield for this process is reported to be directly proportional to the amount of activated carbon employed [234].

Petrier and Luche have reported preliminary results from a study of the reductive properties of an aqueous $Zn—NiCl_2$ (9:1) system [235]. Selective reduction of α,β-unsaturated carbonyl compounds was reported to occur in water/ethanol mixtures. Ultrasound increases the rate of the reaction, but its role is not simply that of an initiator as chemical activation of the zinc, using copper iodide or ammonium chloride, fails to effect reduction of the substrate. In addition, the authors stress that reaction does not occur in the absence of water. Reduction occurs preferentially at the double bond of the enone and is virtually complete before reduction of the carbonyl group begins. This degree of selectivity is not observed in the case of α,β-unsaturated aldehydes.

Further manipulation of the system showed that highly selective reduction of enones could be carried out in the presence of isolated double bonds (Scheme 117). For example, 3 h sonication of carvone (**54**) in the presence of the catalyst gave carvomenthone (**55**). On the other hand, buffering the reaction mixture with NH_3/NH_4Cl or triethylamine gave essentially pure dihydrocarvone (**56**). It appears that this effect is due to selective poisoning of the catalyst rather than the pH of the solution, since the authors report that this effect was only seen in the presence of amino compounds and was absent when other buffering additives were employed. Finally, the reaction under a hydrogen atmosphere with a 1:1 catalyst was examined. Sonication of the reaction mixture gave low rates of reaction and erratic yields. This can be attributed to the degassing effect that accompanies cavitation. By contrast, the stirred reaction is more specific and gave the product resulting from reduction of the unconjugated double bond (**57**), accompanied by a small amount of the fully reduced product (**55**).

From an experimental point of view, the system is similar to that described by Sakai and co-workers [236], who advocate use of precipitated nickel. The principal advantage of the sonochemical approach is that it does not require tedious preformation of the catalyst. In addition, the quantity of catalyst required is reduced. However, the alcoholic co-solvent must be carefully

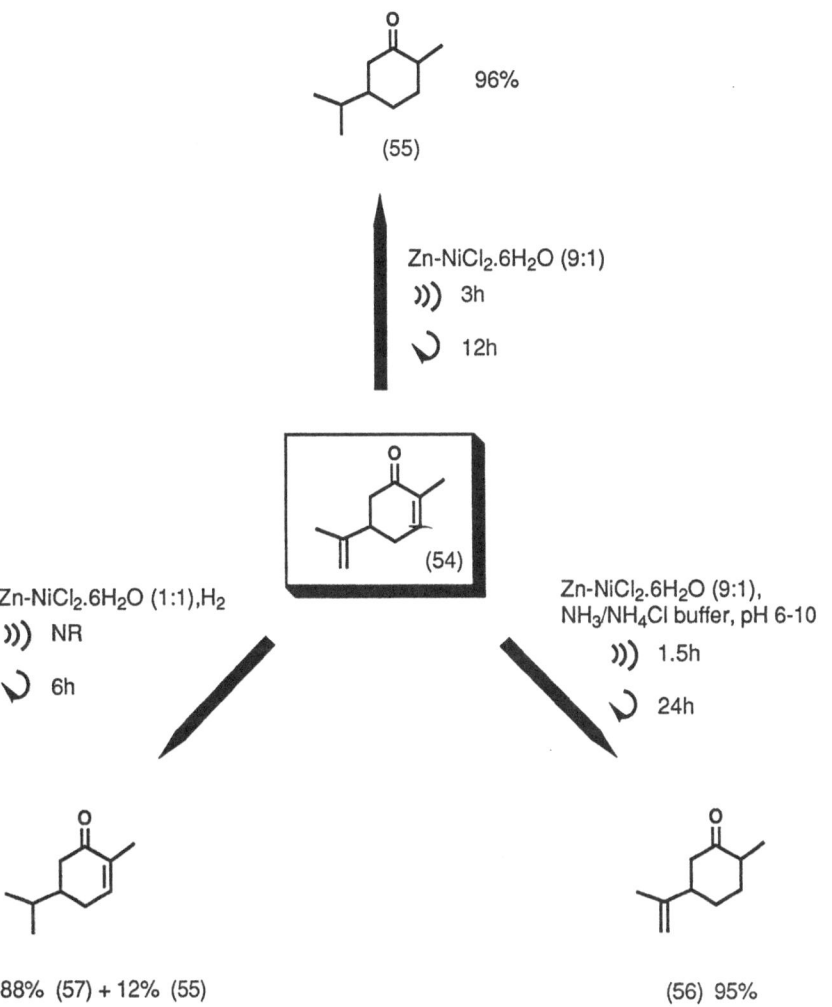

96%

(55)

Zn-NiCl$_2$.6H$_2$O (9:1)
))) 3h
⤸ 12h

(54)

Zn-NiCl$_2$.6H$_2$O (1:1),H$_2$
))) NR
⤸ 6h

Zn-NiCl$_2$.6H$_2$O (9:1),
NH$_3$/NH$_4$Cl buffer, pH 6-10
))) 1.5h
⤸ 24h

88% (57) + 12% (55)

(56) 95%

Scheme 117

chosen to ensure complete solubility of both the starting material and the final product, to allow separation of the two phases.

The mechanism of the process is not currently understood and is complicated by the observation the hydrogen is evolved during the reaction, but does not appear to play a significant role in the reduction process. On the other hand, a performed catalyst (Zn:NiCl$_2$ = 9:1) is completely stable under an inert atmosphere for periods of up to 48 h in the absence of the substrate and reaction is only triggered by addition of the enone.

Sterically hindered amines can usually be oxidised using hydrogen peroxide in the presence of a catalytic amount of Na$_2$WO$_4$ [237]. However, difficulties arise in cases where the amine is attached to a long lipophilic chain. These can be overcome by sonication of the reaction mixture and a series of stable piperidine and piperazine *N*-oxyl radicals were prepared in this way [238] (Schemes 118 and 119). These compounds were subsequently used as spin probes to assess the properties of liquid crystals.

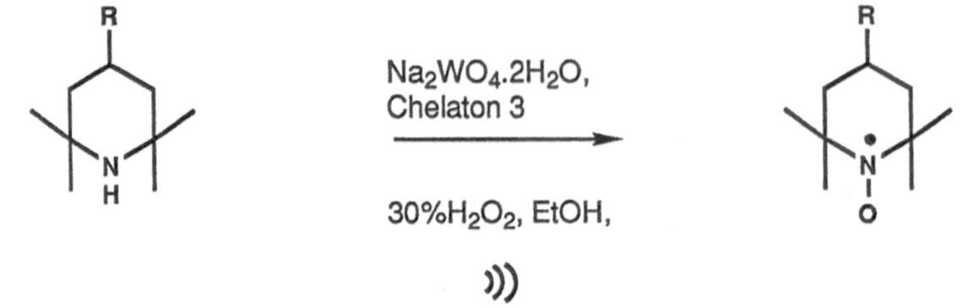

$$Na_2WO_4.2H_2O,$$
Chelaton 3

$$30\%H_2O_2, EtOH,$$

))

R = OCO(CH$_2$)$_{16}$CH$_3$
  = OCOC$_6$H$_4$OC$_8$H$_{17}$-*p*
  = NHCOC$_6$H$_4$OC$_8$H$_{17}$-*p*

Scheme 118

Further reduction occurs in the cases where R=OH or NH$_2$, and the products obtained were the carbonyl or hydroxyimino-*N*-oxyl radicals respectively. The process was equally successful in the case of the diazadispiro [5.1.5.3] systems shown below (Scheme 119). In contrast, standing the reagents together for a period of several weeks only gave trace amounts of the *N*-oxyl radicals.

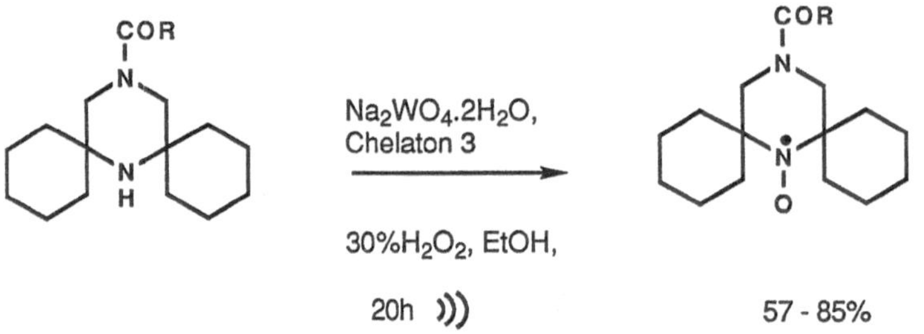

$$Na_2WO_4.2H_2O,$$
Chelaton 3

$$30\%H_2O_2, EtOH,$$

20h))

57 - 85%

R = *p*-C$_6$H$_4$OC$_8$H$_{17}$
  = (CH$_2$)$_{16}$CH$_3$
  =C$_5$H$_{11}$

Scheme 119

13 The Effects of Ultrasound
on Transition Metal Catalysts

Investigation of a gas phase reaction suggested that the effects of ultrasound on heterogeneous catalysts were simply due to increased mass transport [239]. However, it has also been clearly shown that palladium and platinum blacks prepared by reduction of the metal salts in the presence of formaldehyde had up to 30% greater surface areas and showed increased activity in a number of representative reactions. These included the decomposition of water, the hydrogenation of hex-1-ene, and the oxidation of ethanol [240]. In addition, the catalysts showed increased paramagnetism and appeared to contain a higher concentration of atomic metal. Interestingly, Mal'tsev observed that increasing the frequency of the ultrasound used in the preparation of platinum blacks from 20 kHz to 3 MHz resulted in an increased level of catalytic activity. Conversely, the opposite trend has been observed in the case of the palladium blacks. There is no obvious explanation for this. A similar technique was used to prepare suspensions of platinum on silica gel. Hence, reduction of solutions of platinum complexes in the presence of ultrasound (440 kHz, 5 W cm^{-2}) increased the available surface area of platinum by 80% with respect to a control sample. Further evidence has been provided by Cioffi and Prestegard who have suggested that modification of the catalyst surface may also be a factor to consider. Raney nickel was used to catalyse the incorporation of deuterium into a glycosphingolipid [242]. Catalytic deuteration of non-reducing monosaccharides using this catalyst in D_2O was demonstrated some time ago. However, the forcing conditions required restricted the general applicability of the method to simple saccharides. It was then shown that sonication of the substrate and deuterated Raney nickel in D_2O gave unrearranged products in virtually quantitative yield. Moreover, the selectivity of the process appeared to be better than that observed under thermal conditions. The authors report that preliminary studies showed that similar results were obtained if the catalyst was exposed to ultrasound before addition of the substrate. This suggests that the rate enhancements observed may well owe more to modification of the catalyst surface than a simple increase in mass transport.

Some Bulgarian workers have examined the effects of ultrasound on the properties of Cr—Mo—O catalysts used in the oxidation of methanol. However, the results of these investigations are unclear as the two publications are directly contradictory [243, 244]!

Low valent nickel complexes catalyse the homo-coupling of aryl halides. The required complexes can be generated in situ from a mixture of a nickel (II) salt, triphenylphosphine, potassium iodide and zinc powder [245]. Interestingly, sonication of this complex mixture allows aryl sulphonates to be used in place of the conventional halides and represents one of the few cases where aryl-oxygen bond fission occurs [246]. The authors remark that the reactions typically took 4 h at 60 °C and represented a 20–50% increase over that of the stirred reaction (Scheme 120). It would be interesting to know whether this margin could be increased if the reaction temperature was decreased in view of the proposals of the "hot spot" theory.

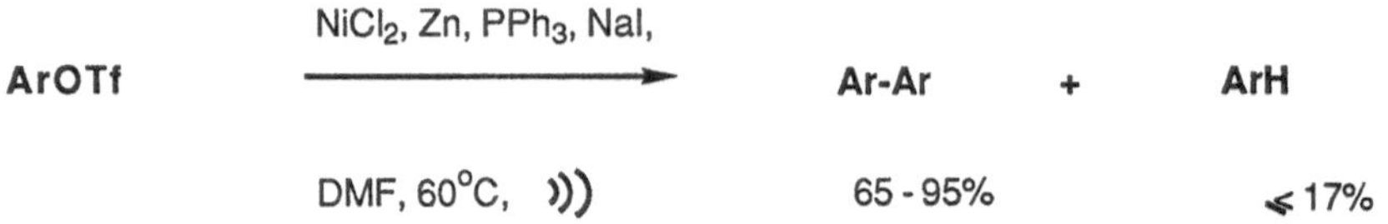

$$\text{ArOTf} \xrightarrow[\text{DMF, 60°C, }))]{\text{NiCl}_2\text{, Zn, PPh}_3\text{, NaI,}} \text{Ar-Ar} + \text{ArH}$$

Ar-Ar 65 - 95%   ArH ⩽17%

Scheme 120

Palladium on charcoal is commonly used to promote cleavage of benzyl ether protecting groups and ultrasound was successfully used to enhance the rate of hydrogenation of the nocardicin derivative (**58**) [247] (Scheme 121)

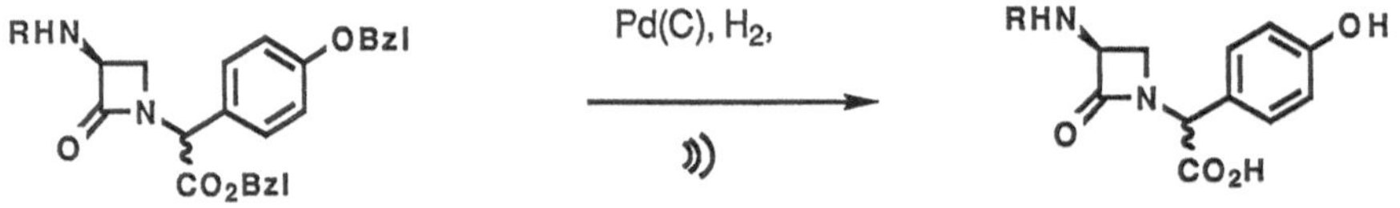

Scheme 121

Furthermore, Han and Boudjouk demonstrated that formic acid could be used as a source of hydrogen in the palladium catalysed reduction of olefins [248]. Similarly, hydrazine can also be used although the reaction carried out in absolute ethanol is marginally faster at reflux than when carried out in a laboratory cleaning bath [249]. Platinum-catalysed hydrosilylation reactions are also promoted by ultrasound and high yields of the desired product can be obtained within extremely short periods of time [250]. However, rate increases are typically less than ten-fold.

Ultrasound has also been shown to play a role in a number of industrially important processes. These include the liquefaction of coal by hydrogenation with Cu/Zn [251], ammonia synthesis [252], and a number of polymerization reactions [253].

14 Transition Metal Carbonyls and Ultrasound

π-Allyltricarbonyliron lactone complexes are useful precursors for organic synthesis. They were first reported in 1964 [254] and have since been shown to be available from a variety of substrates [255]. For example, they may be prepared from alkenyl epoxides or various butenediols [256] and their derivatives by treatment with tetracarbonyl iron [257]. Work in our laboratories had shown that these were useful precursors for a wide range of naturally occurring β and δ-lactones and lactams [258] (Scheme 123).

The tetracarbonyliron species had originally been generated by photolysis of iron pentacarbonyl. However, the hazardous nature of this reagent prompted us to examine other methods for its generation. Experiments showed

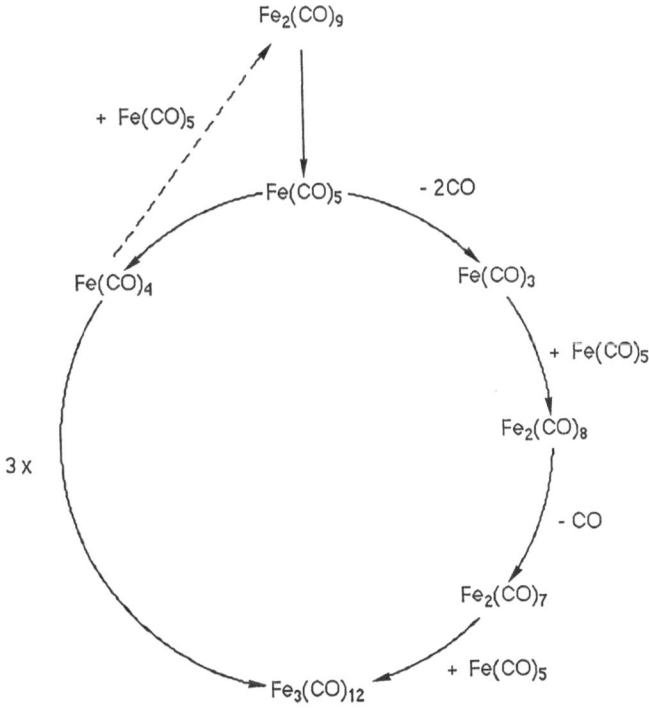

Fig. 23. Proposed mechanism for the formation of $Fe_3(CO)_{12}$ on sonolysis of benzene solutions of $Fe(CO)_5$ and $Fe_2(CO)_9$

that ferrilactones were readily available as a result of treating vinyl epoxides with diiron nonacarbonyl in THF solution. Diiron nonacarbonyl itself is partially soluble in THF and IR studies clearly show the presence of $Fe(CO)_5$ in solution accompanied by "at least one other carbonyl species" [259]. Cotton has suggested that coordination of a THF molecule to the vacant site of the iron may be occurring [260], indicating that the active species is in fact $Fe(CO)_4 \cdot THF$. Aumann has also invoked the intermediacy of a tetracarbonyliron species to explain the results of the photolytic reaction [257]. This is supported by the availability of diiron nonacarbonyl on photolysis of iron pentacarbonyl which probably occurs by reaction of a tetracarbonyliron species with excess starting material.

It was also discovered that sonolysis of slurries of diiron nonacarbonyl and vinyl epoxide in benzene solution gives good yields of the required product [261]. However, Suslick reports that the only products generated by sonolysis of $Fe_2(CO)_9$ are $Fe(CO)_5$ and finely divided iron [262]. Sonolysis of $Fe(CO)_5$ itself leads to formation of the dark green $Fe_3(CO)_{12}$. In considering the mechanism of this reaction Suslick concluded that the initial step was formation of a *tri*-carbonyliron intermediate. Subsequent reaction with $Fe(CO)_5$ could then be invoked to explain the presence of $Fe_3(CO)_{12}$ (Figure 23).

The intermediacy of $Fe(CO)_4$ was ruled out on the grounds that it would be trapped by excess $Fe(CO)_5$ as soon as it was formed, producing $Fe_2(CO)_9$. Diiron nonacarbonyl was not observed as a product of this reaction and despite his observation that sonochemical cleavage of this species occurred as fast as that of $Fe_3(CO)_{12}$ production, Suslick emphatically concluded that

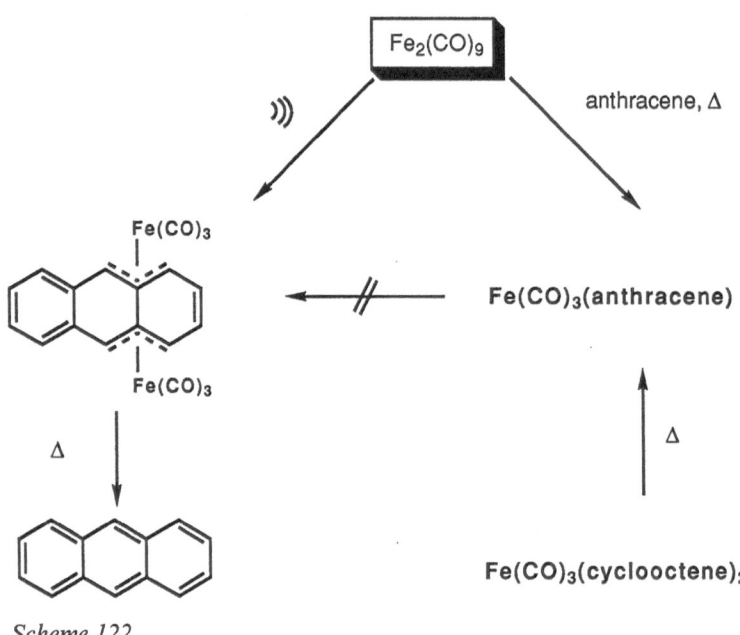

Scheme 122

Table 8

Substrate	Ferrilactone	% Yield	
		Fe₂(CO)₉/THF	Fe₂(CO)₉ /))) PhH
	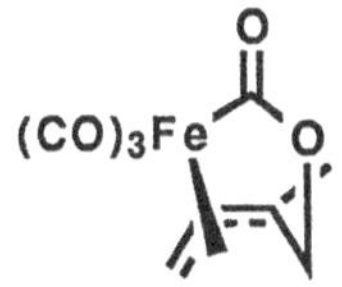	75	90
	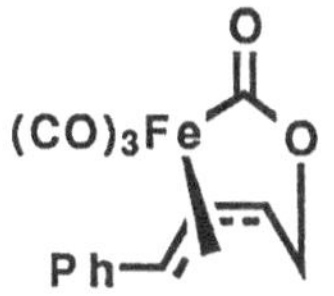	100	68
	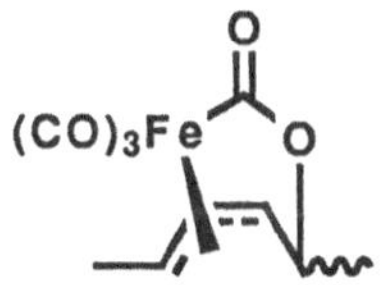	64	55
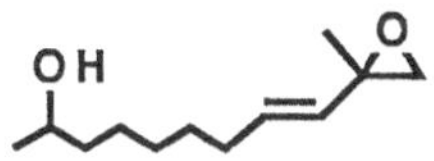	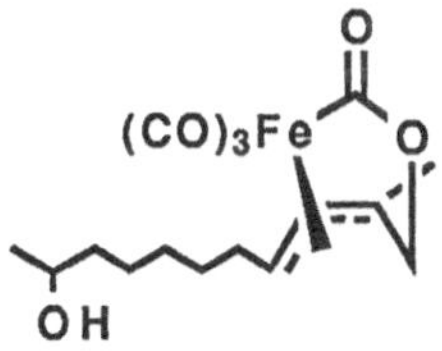	91	91
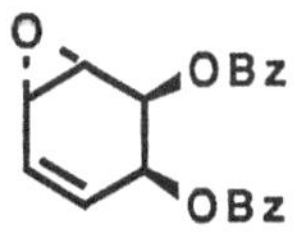 (59)	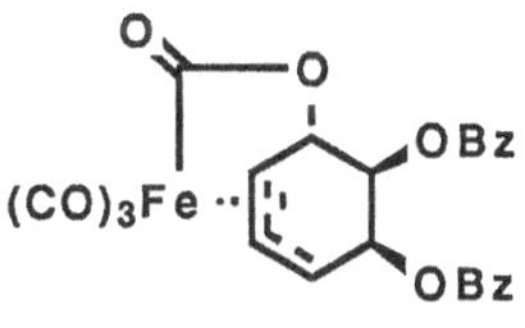	14*	100*
	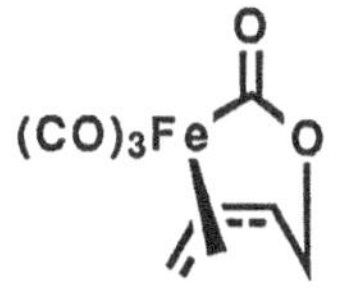	73	55

* 4 equivalents of Fe₂(CO)₉ used

107

Fe(CO)$_4$ was not an intermediate on the pathway to Fe$_3$(CO)$_{12}$. However, he later conceded that the presence of Fe(CO)$_4$ species could not be ruled out [3].

The situation is further complicated by reports that sonolysis of anthracene and Fe$_2$(CO)$_9$ in hexane gives Fe$_3$(CO)$_{12}$ accompanied by a novel diiron-anthracene complex [263]. This product is not available using standard chemistry as the thermal reaction gives the mono-adduct as the sole product, which cannot be subsequently converted to the diiron complex (Scheme 122). This suggests that the reaction goes via an Fe$_2$(CO)$_n$ species and not by sequential complexation.

Nevertheless, the two methods developed for formation of ferrilactone complexes show wider substrate tolerance than either the thermal or photolytic reactions and avoid problems that can occur at higher temperature, such as decarbonylation, decarboxylation or metal-catalysed hydrogen shifts [265]. They both give comparable yields of ferrilactone under the conditions described (see Table 8). The one exception is the case of vinyl epoxide (**59**), when four equivalents of Fe$_2$(CO)$_9$ were required [206]. This is probably due to coordination of the reactive species to the oxygens of the *cis*-dibenzoyl system reducing the effective concentration of tetracarbonyliron in solution. Conversely, the ultrasound reaction presumably takes place at the surface of the solid Fe$_2$(CO)$_9$ where the *local* concentration of Fe(CO)$_4$ is much higher and this is reflected in the yields obtained.

We have not been able to reproduce this reaction using Fe(CO)$_5$ in our laboratories. However, this is probably due to the low intensities of ultra-

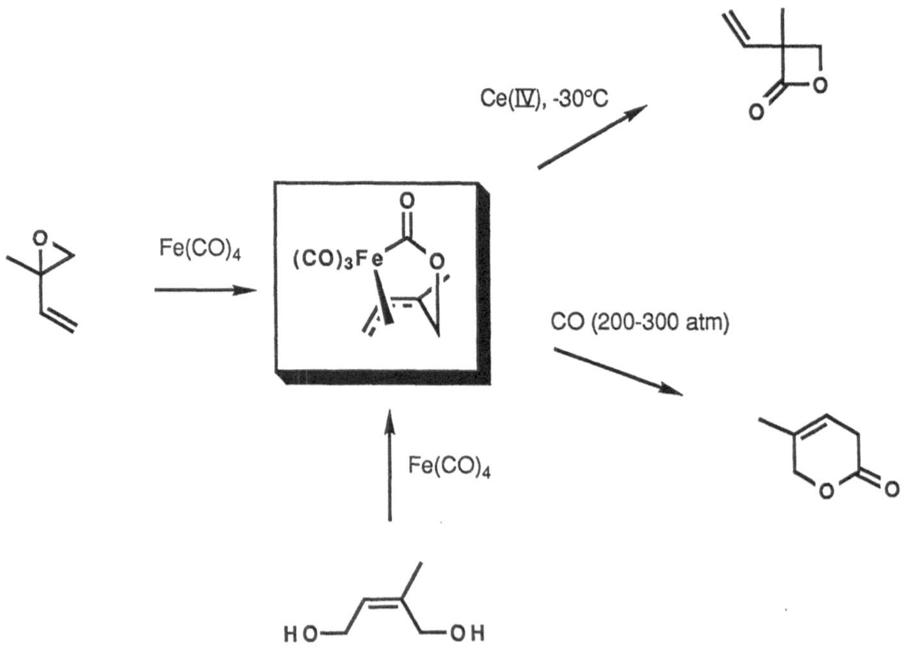

Scheme 123

sound obtainable using a cleaning bath. Whilst this is clearly sufficient to produce cavitation in heterogeneous slurries of $Fe_2(CO)_9$, is inadequate for the corresponding homogeneous reaction for reasons already discussed. Suslick has shown that $Fe(CO)_5$ will undergo substitution by phosphines under more intense irradiation [266], although the sequential loss of carbonyl ligands is not a feature of either the thermal or photolytic reactions.

The ferrilactone complexes are obtained as air stable, crystalline solids that can be isolated using standard organic techniques, including silica gel chromatography, and have been shown to be versatile intermediates for the synthesis of lactones. For example, low temperature oxidation with cerium (IV) gives good yields of the β-lactones arising from coupling of the lactone carbonyl to C2 of the π-allyl unit [267]. That is, the overall transformation is equivalent to formal addition of carbon monoxide across the oxirane ring. Conversely, exhaustive carbonylation of these complexes produces the δ-lactones to the exclusion of any other product [257]. Hence, this provides access to two complimentary processes for the preparation of either β- or δ-lactones (Scheme 123).

Indeed, this provides a rapid entry to α,β-unsaturated lactone natural products such as massoialactone and parasorbic acid [268]. Alternatively, carbonylation at lower temperatures (90 °C) and high CO pressures (200 to

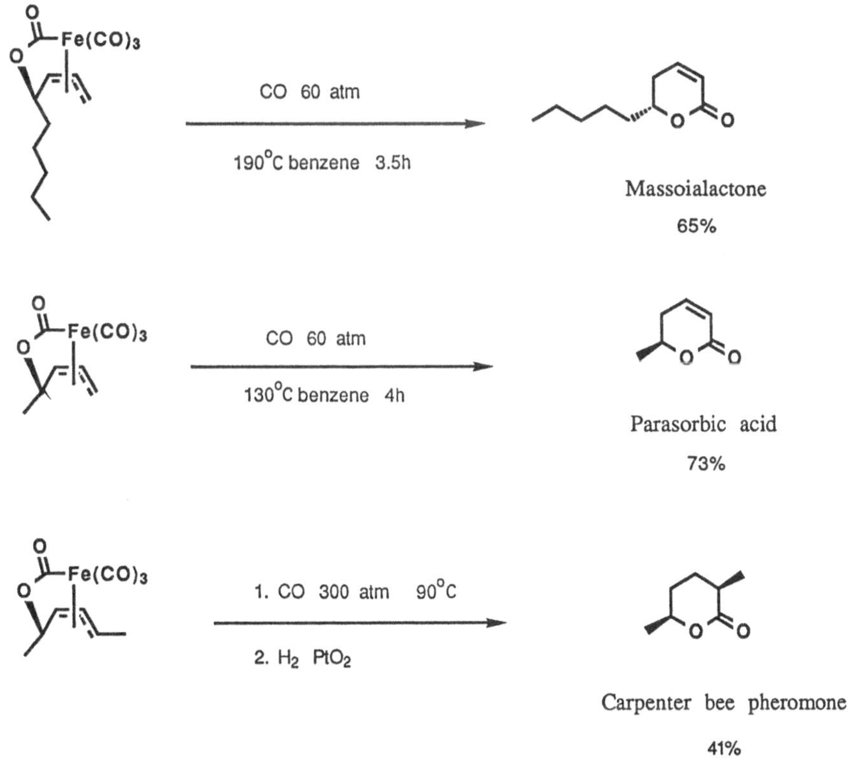

CO 60 atm

190°C benzene 3.5h

Massoialactone

65%

CO 60 atm

130°C benzene 4h

Parasorbic acid

73%

1. CO 300 atm 90°C

2. H₂ PtO₂

Carpenter bee pheromone

41%

Scheme 124

300 atm.) affords β,γ-unsaturated lactones, which give the carpenter bee pheromone and the antibiotic malyngolide on hydrogenation (Schemes 124 and 125).

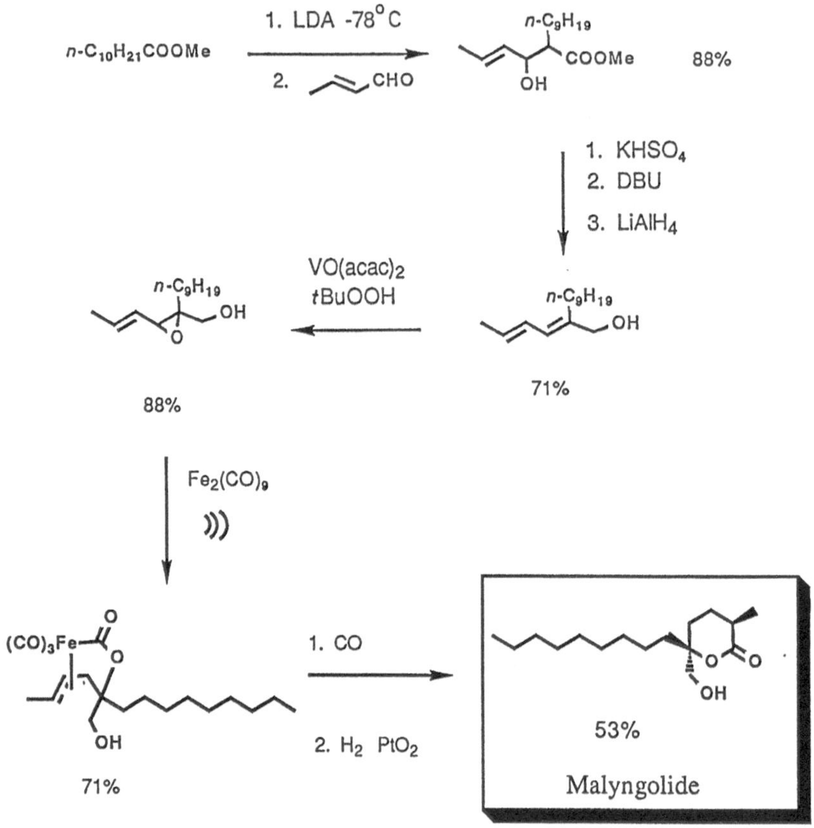

Scheme 125

Use of the Sharpless epoxidation allows access to optically active alkenyl epoxides and both the reaction with diiron nonacarbonyl under sonolytic conditions and the subsequent carbonylation proceed with retention of configuration, thus allowing us to predetermine the stereochemistry at of the δ-lactone at the C5-centre (Scheme 125).

This methodology was exploited in the synthesis of key structural fragments of the ionophere antibiotic CP 61 405 [269, 270] (Scheme 126) and the spiroacetal portion of the potent antiparasitic agents, the avermectins [270] (Scheme 127).

Ferrilactones are not only useful precursors for the synthesis of β- and δ-lactones. Reaction with primary amines in the presence of a Lewis acid catalyst, such as Et_2AlCl or $ZnCl_2 \cdot TMEDA$, gives the corresponding ferrilactam complexes. Low temperature oxidation with cerium (IV) then gives access to the monocyclic β-lactams in an analogous manner [271] (Scheme 128).

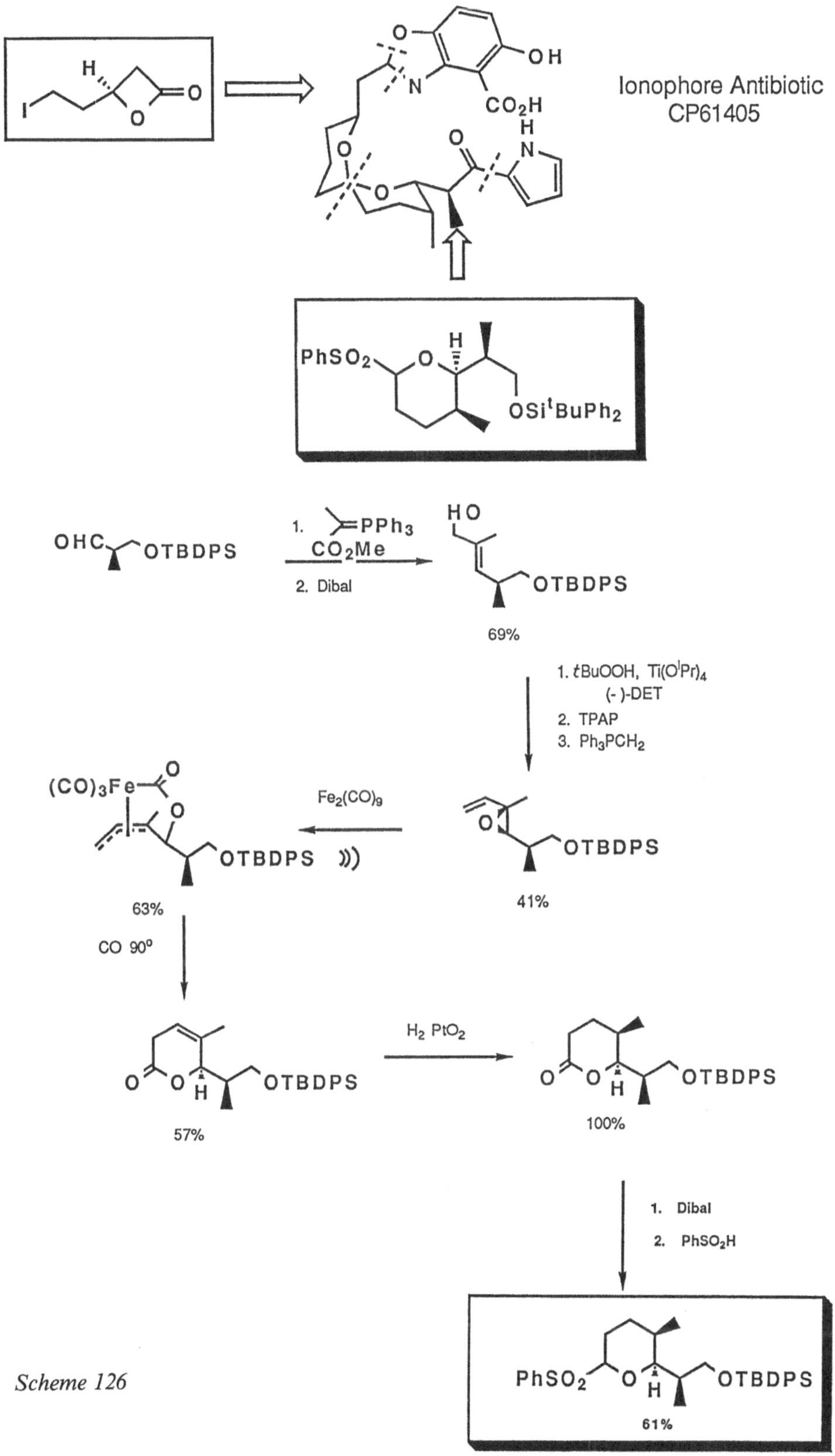

Ionophore Antibiotic
CP61405

OHC⟍OTBDPS

1. ⟍=PPh₃
   CO₂Me
2. Dibal

→ 69%

HO⟍OTBDPS

1. tBuOOH, Ti(O^{i}Pr)₄
   (-)-DET
2. TPAP
3. Ph₃PCH₂

41%

Fe₂(CO)₉
)))

(CO)₃Fe⟍O
63%

CO 90°

57%

H₂ PtO₂

100%

1. Dibal
2. PhSO₂H

PhSO₂⟍O⟍OTBDPS
61%

Scheme 126

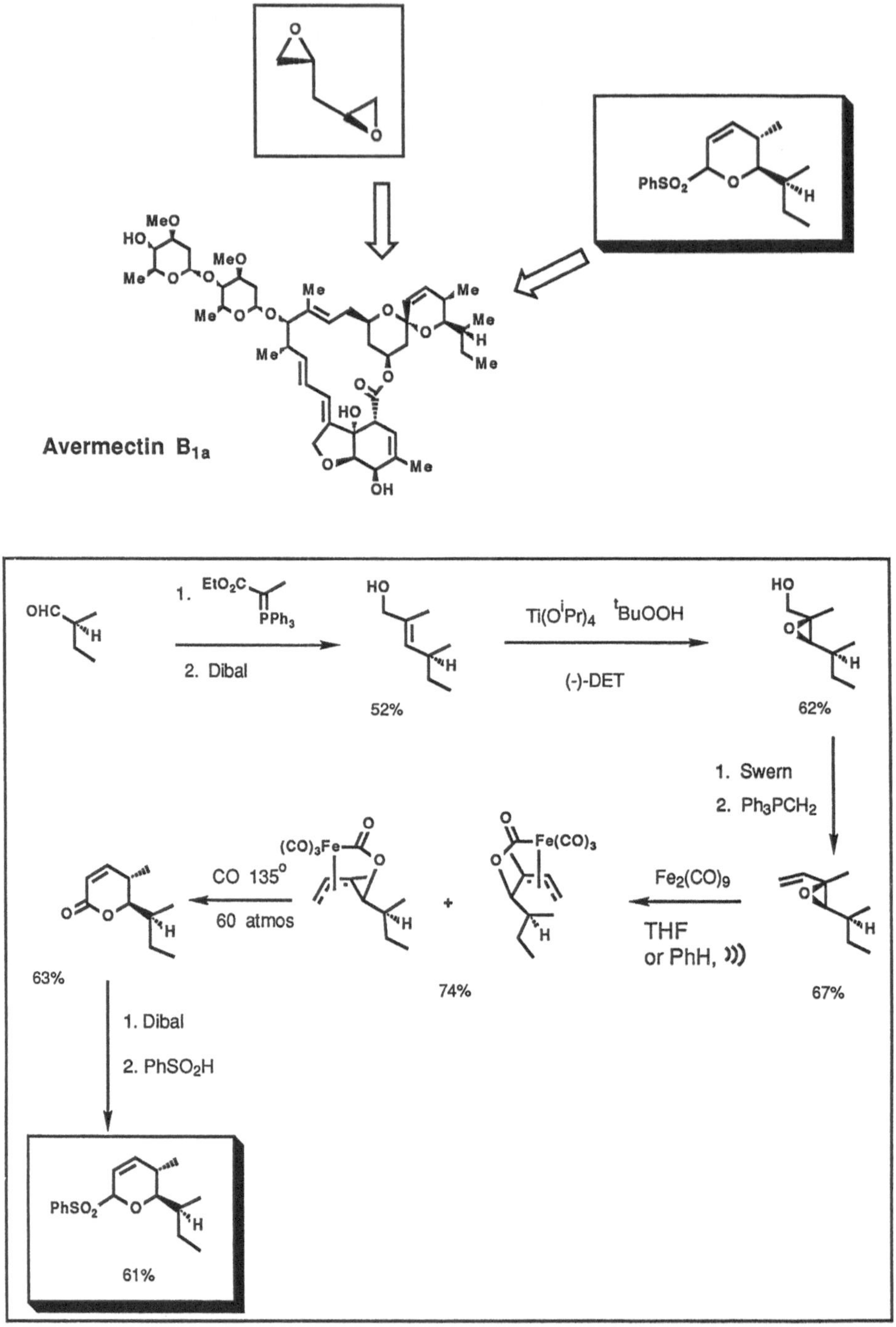

Avermectin B1a
1. EtO2C PPh3
2. Dibal
52%
Ti(OiPr)4 tBuOOH
(-)-DET
62%
1. Swern
2. Ph3PCH2
Fe2(CO)9
THF
or PhH,)))
67%
74%
CO 135°
60 atmos
63%
1. Dibal
2. PhSO2H
61%

Scheme 127

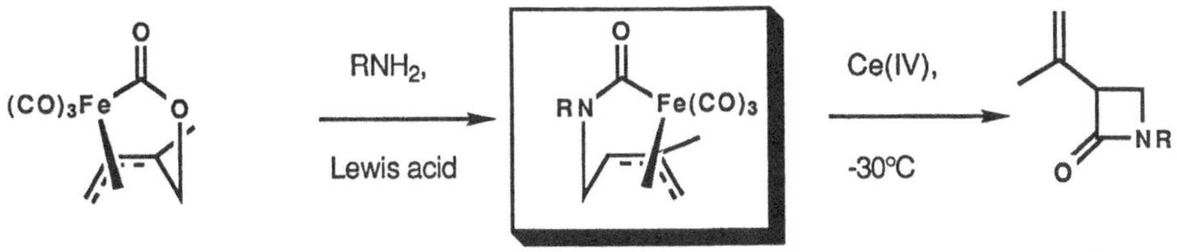

Scheme 128

The products of this reaction are particularly useful in that they possess unsaturated side chains at C3. That is, they are the regioisomers of the products that can be obtained from the widely used reaction of dienes and chloro-sulphonylisocyanate (CSI) [272] (Scheme 139).

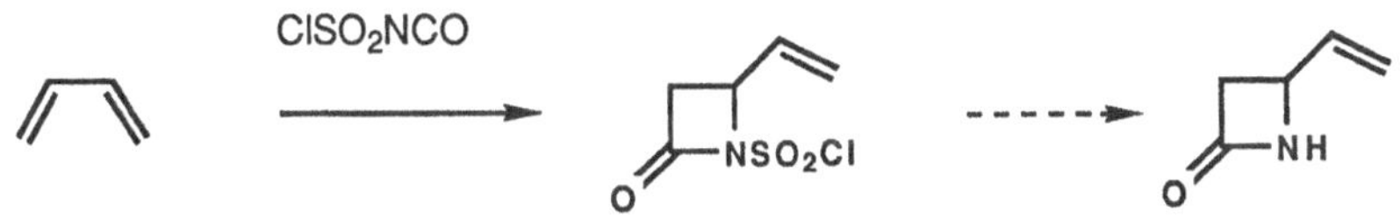

Scheme 129

Further examination of the products of the organoiron reaction showed that the route was generally applicable to the synthesis of a range of pharmaceutically important β-lactam antibiotics which included thienemycin [273] (Scheme 130) and the monocyclic β-lactam systems of the nocardicins [274] (Scheme 131) and monobactams.

Further studies have shown that $Fe(CO)_4$ species are readily trapped by 1,3-conjugated dienes and the η^4-(diene)$Fe(CO)_3$ complexes are isolated in virtually quantitative yield [277]. In this case, reaction is known to proceed via initial formation of the η^2-(olefin)-$Fe(CO)_4$ complex. Loss of carbon monoxide then gives the η^4-(diene)-$Fe(CO)_3$ complex [275] (Scheme 132).

This reaction is of great synthetic use in the light of the moderating effect of the $Fe(CO)_3$ unit on the reactivity of the unsaturated system towards hydrogenation, electrophilic attack and Diels-Alder reaction. Taken in conjunction with the number of methods in existence for the recovery of the diene these results suggest that the tricarbonyliron moiety may be a useful protecting group for conjugated dienes. In addition, its steric bulk can be used to control the stereochemical outcome of the reaction, and hydride abstraction allows access to the wide range of chemistry associated with the tricarbonyldienylium cation [276]. However, the major problem to date has been that yields of the both the thermal and photolytic reaction between the diene and $Fe(CO)_5$ are extremely modest. This obviously constitutes a barrier to the general acceptance of these species as synthetic intermediates which has yet to be overcome.

Problems arising from the sensitivity of both products and starting materials to aerial oxidation dictate that reactions must be carried out under inert atmospheres. The most commonly used procedure involves heating a mixture of the diene and $Fe(CO)_5$ in an high boiling inert solvent, such as di-*n*-butyl

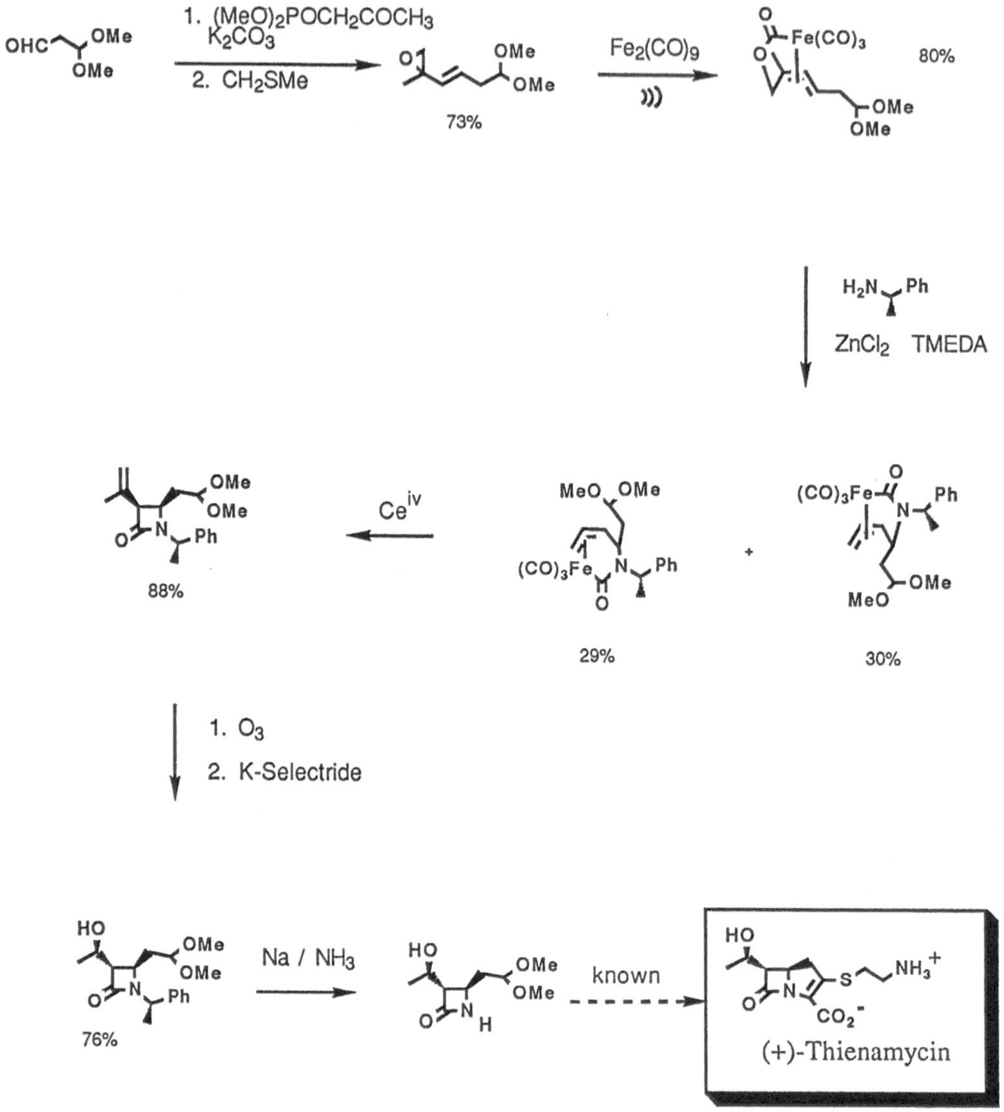

Scheme 130

ether, for long periods of time. However, the sensitivity of the complex to peroxides present in these solvents make it imperative to filter the solvent through basic alumina prior to use. A number of other modifications have also been suggested, and yet final yields of complex remain disappointingly low.

Against this background, investigations into the synthesis of such compounds by reaction of a conjugated diene with $Fe_2(CO)_9$ under the sonolytic conditions used for the generation of ferrilactone complexes was begun. In the initial attempt, reaction with 1-acetoxybuta-1,3-diene in benzene afforded

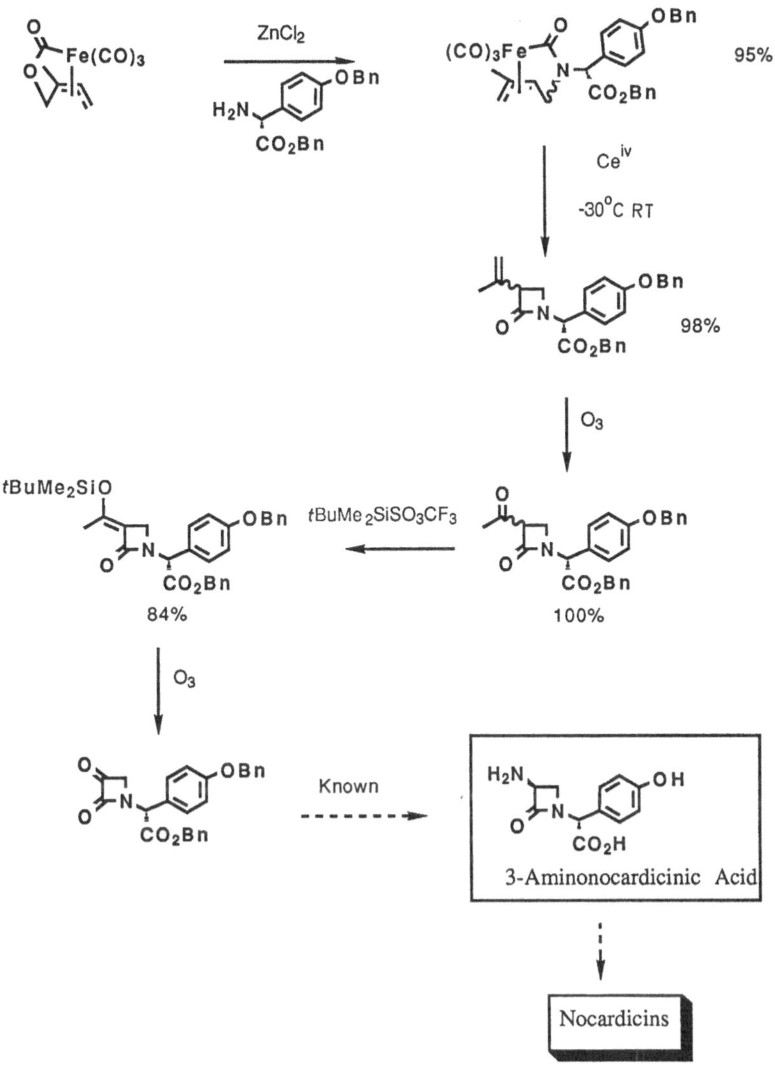

Scheme 131

a quantitative yield of the η^4-(diene)tricarbonyliron complex. Interestingly, this corresponds to the 15% yield obtained when the reaction was carried out using $Fe_2(CO)_9$ in THF solution. This is in sharp contrast to its efficacy as a source of tetracarbonyliron in the preparation of ferrilactone complexes from vinyl epoxides.

Extension of this methodology showed that the method was generally applicable to a variety of conjugated dienes and tolerant of both electron-donating and withdrawing groups [277] (Table 9). At this point it should be stressed that the reactions were carried out at ambient temperature, without the exclusion of air and without prior purification or drying of the solvent. Despite

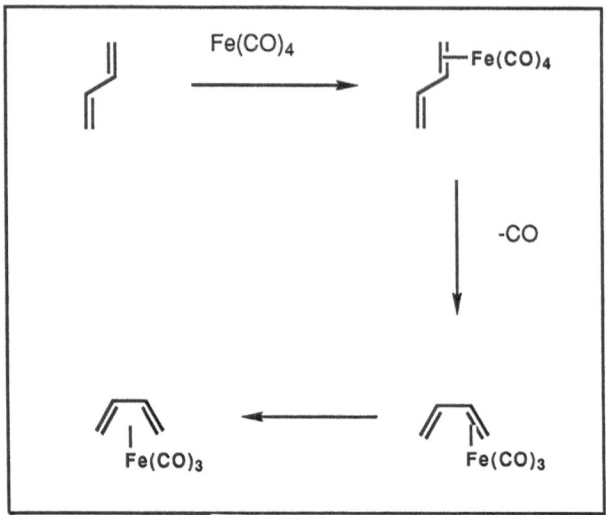

Scheme 132

Table 9

Diene	Complex	% Yield
![diene OAc]	![complex OAc Fe(CO)₃]	100%
CO₂Me diene	CO₂Me Fe(CO)₃	100%
OH diene	OH Fe(CO)₃	51%
OBz OBz diene	(CO)₃Fe OBz OBz	75%
diene	Fe(CO)₃	95%

this, the unoptimised yields of complex obtained after 1 h of sonolysis are consistently higher than previous literature preparations.

Reaction with more complex polyenes showed a number of interesting features. For example, reaction with β-myrcene gives a 95 % yield of the product shown. That is, reaction occurs specifically with the conjugated diene and the product isolated is uncontaminated by species arising from complexation of the isolated double bond.

Reaction with pseudionone is of particular interest in that it was possible to isolate a 75 % yield of the kinetic product after 1 h of sonolysis. Further exposure to ultrasound, or standing at room temperature resulted in isomerisation to give the 5,9-diene complex (Scheme 133). Hence, variation of the reaction time allows selective access to good yields of both isomers. This is not possible by conventional methods and reflects the mildness of the conditions used.

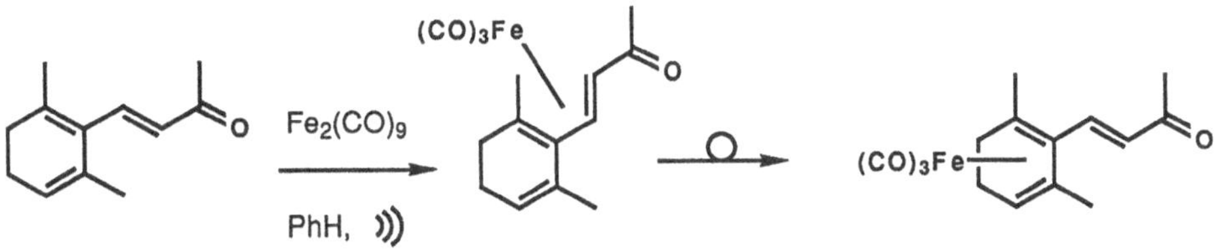

Scheme 133

In the analogous reaction of β-ionone with $Fe_2(CO)_9$, 1 h of sonolysis gave a 4:1 mixture of endo and exo complexes (Scheme 134). In contrast, Cais and Maoz report that the thermal reaction of the diene with $Fe(CO)_5$ leads to isolation of a 3:1 ratio of products in favour of the exo-isomer [278]. However, it should also be noted that the overall yield of the thermal reaction was only 21 %, that is a factor of three less than that obtained under sonolytic conditions.

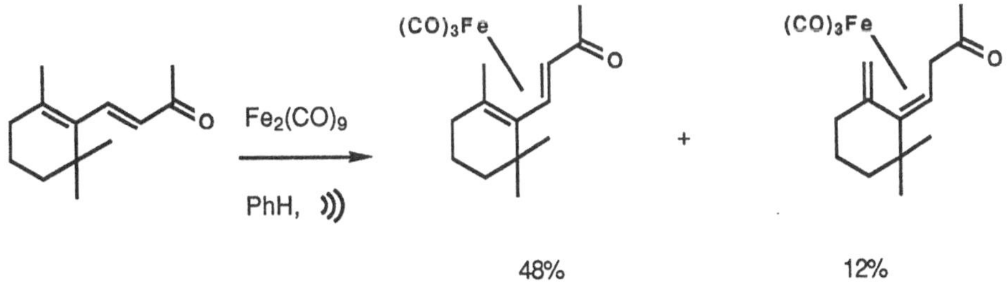

Scheme 134

Similarly, it has been demonstrated that trimethylenemethane tricarbonyliron (60) is readily available by sonolysis of equimolar amounts of $Fe_2(CO)_9$ and 3-chloro-2-chloro-methylprop-1-ene in 30/40° petroleum ether. This gives a remarkable 90 % yield of the desired product [277] (Scheme 135).

In contrast, the highest recorded yield of the thermal reaction with disodium tetracarbonylferrate was a meagre 32%, described by Emerson et al. [279].

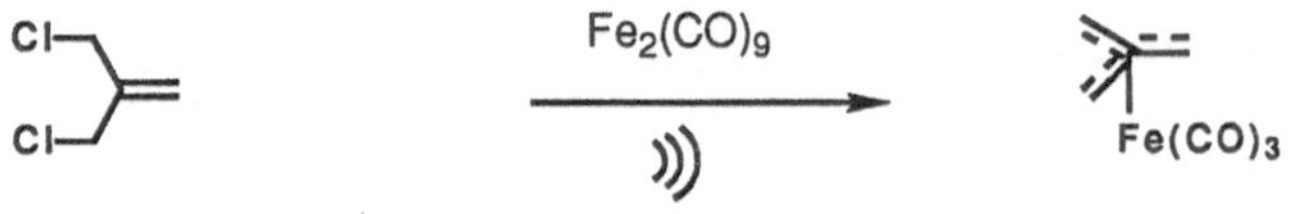

(60)

Scheme 135

Interestingly, attempts to use the bis-allylic alcohol as the substrate in the analogous reaction led to isolation of the ferrilactone complex (61) as the major product, accompanied by a trace of the trimethylenemethane complex. Conversely, reaction in THF at room temperature resulted in a complete reversal of the product distribution [256] (Scheme 136). The reasons for this are currently unclear.

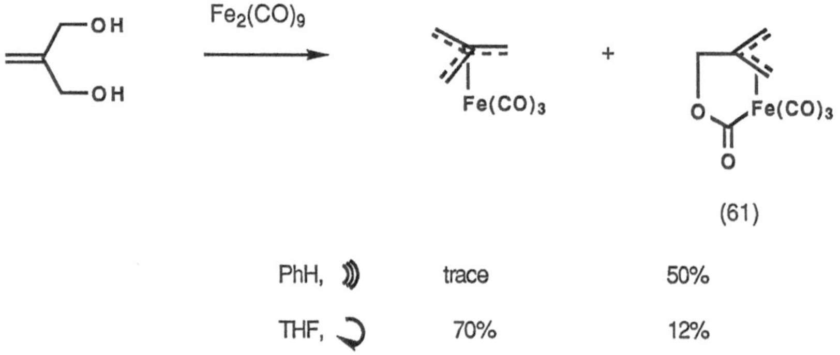

(61)

PhH,)))	trace		50%
THF,	70%		12%

Scheme 136

Suslick has made an extensive study of the sonochemistry of $Fe(CO)_5$ which he has used as a probe to explore the chemical effects of high intensity ultrasound. Suslick and Johnson [280] have also shown that sonication greatly facilitates the preparation of early transition metal carbonyl anions. Hence, sonication of vanadium trichloride and sodium sand in THF solution gave a 35% yield of $NaV(CO)_6$ under 4.4 atmospheres of carbon monoxide at 10 °C. The equivalent thermal reaction requires the reaction to be caried out at 160 °C under 200 atmospheres of carbon monoxide. That is, the temperatures and pressures produced by cavitation are comparable to the bomb conditions normally required for the preparation of these compounds [281]. Suslick's review [3] presents further evidence in support of this original observation; however, no further details of this work have appeared to date.

A variety of coordinatively unsaturated metal carbonyl species are known to be potent homogeneous catalysts. Sonication of 1-pentene in the presence of $Fe(CO)_5$ produces the thermodynamic mixture of *cis* and *trans* pent-2-ene [282]. Under these conditions the reaction is 105 times faster than the equivalent thermal reaction [262]. The increase is less marked in the case of sterically hindered alkenes, e.g. 2-ethylpent-1-ene, and does not occur in the absence of β-hydrogens. This is consistent with the reaction occurring via a hydrido-π-allyl intermediate and FT IR spectroscopy implies the presence of $Fe(CO)_4$(pentene) under these conditions. The examples given demonstrate the ease with which coordinatively unsaturated transition metal carbonyl species can be generated. The specificity of the reactions carried out at, or slightly below, ambient temperature contrasts with the low yields of products obtained via equivalent thermal reactions and clearly shows potential for the preparation of a wide variety of organometallic complexes.

15 Conclusion

In conclusion, the beneficial effects of ultrasound on reactions in non-aqueous solvents have only been appreciated within the past few years and a great deal more exploration is required before their potential for application to synthesis is fully realised, particularly on a large preparative scale.

One of the criticisms that has been levelled is that ultrasound simply provides a new way of speeding up old reactions and has not produced any new chemistry. However, this is not the case and the sheer variety of reactions detailed in this review demonstrates the relevance of this work to synthetic chemists. Furthermore, the current level of interest in this topic is increasing at such a rate that the widespread acceptance of ultrasound as a synthetic tool is only a matter of time.

16 References

1. Richards WT, Loomis AL (1927) J. Am. Chem. Soc. 49: 3086
2. Wood RW, Loomis AL (1927) Phil. Mag., Ser. 7 4: 417
3. Suslick KS (1986) Ultrasound in synthesis. In: Scheffold R (ed) Modern synthetic methods, Springer, Berlin Heidelberg New York, vol 4 p 1
4. a) Luche J-L (1982) L'actualite Chimique 21; b) Boudjouk P (1983) Nachr. Chem. Tech. Lab. 31: 78; c) Mason TJ (1984) Lab. Prac. 33: 13; d) Mason TJ (1986) Ultrasonics 24: 245; e) Bremner D (1986) Chem. in Brit. 22: 633; f) Suslick KS (1986) Adv. Organomet. Chem. 25: 73; g) RSC Sonochemistry Symposium, University of Warwick, 1986, Ultrasonics 25:1 (1986); h) Lorimer JP, Mason TJ (1987) Chem. Soc. Rev. 16: 239; i) Lindley J, Mason TJ (1987) Chem. Soc. Rev. 16: 275; j) Suslick KS (1989) Scientific American 260: 62
5. Moon S, Duchin L, Cooney JV (1979) Tetrahedron Lett. 19: 3917
6. Han B, Boudjouk P (1982) Tetrahedron Lett. 23: 1643
7. Lukevics E, Gevorgyan VN, Goldberg YS (1984) Tetrahedron Lett. 25: 1415
8. a) Lauterborn W, Hentschel W (1986) Ultrasonics 24: 59; Lauterborn W (1982) Appl. Sci. Res. 38: 165; b) Lauterborn W, Bjorn L (ed) (1984) Finite-amplitude wave effects in fluids, Proc. 1973 Symp. IPC Science and Technology, Guildford, p 195
9. a) Lord Rayleigh (1917) Philos. Mag. 34: 94; b) Crum LA (1982) Appl. Sci. Res. 38: 101; c) Neppiras EA (1980) Physics Rep. 61: 159; d) Coakley WT, Nyborg WL (1978) In: Fry FJ (ed) Ultrasound: Its use in medicine and biology, Elsevier, New York, part 1 p 77; e) Apfel RE (1981) In: Edmonds PD (ed) Methods of experimental physics: Ultrasonics, Academic, New York, vol 19 p 356
10. Fujikawa S, Akamatsu T (1980) J. Fluid Mech. 97: 481
11. Margulis MA, Dmitrieva AF (1982) Zh. Fiz. Khim. 56: 323
12. a) Seghal C, Steer RP, Sutherland RG, Verral RE (1979) J. Chem. Phys. 70: 2242; b) Seghal C, Sutherland RG, Verral RE (1980) J. Phys. Chem. 84: 396
13. Nolting BE, Neppiras EA (1950) Proc. Phys. Soc. B63: 674
14. a) Frenzel H, Schultes H (1935) Z. Phys. Chem. B. 27: 42; b) Frenzel J (1940) Acta Physicochim. (USSR) 12: 3; c) Bresler S (1940) Acta Physicochim. 12: 323
15. Frenkel YI (1940) Zh. Fiz. Khim. 12: 305
16. Schulz R, Henglein A (1953) Z. Naturforsch 8b: 160
17. Henglein A, Mohrhauer H (1958) Z. Phys. Chem. Neue Folge 18: 43
18. Henglein A, Fischer C-H (1984) Ber. Bunsenger, Phys. Chem. 88: 1196
19. Niemczewski B (1980) Ultrasonics 18: 107
20. Suslick KS, Gawienowski JJ, Schubert PF, Wang HH (1984) Ultrasonics 33: 21
21. Basedow AM, Ebert KH (1977) Adv. Polym. Sci. 22: 83
22. Margulis M, Grundel LM (1982) Zh. Fiz. Khim. 56: 1445, 1941 and 2592
23. Sonochemistry Group and Luche J-L (1987) Ultrasonics 25: 40

24. Akulicher VA, Sirotyuk MG, Rozenberg LD (1971) In: Rozenberg LD (ed) High intensity ultrasonic fields, Plenum, New York, p 203
25. a) Schumb WC, Peters H, Mulligan LH (1937) Metals and Alloys 5: 126; b) Blake FG (1949) Phys. Rev. 75: 1313; c) Horton JP (1953) Acoust, Soc. Am. 25: 480; d) Connolly W, Fox FE (1954) J. Acoust. Soc. Am. 26: 943; e) Bebchuck AS (1957) Akust. Zhur. 3: 90; f) Rozenberg LD (1960) Ultrasonic News 4: 4
26. a) Chendke PK, Fogler HS (1974) Chem. Engin. J. 8: 165; b) Weisser A (1953) J. Acoust. Soc. Am. 25: 651; c) Bronskaya LM, Vigderman VS, Sokol'skaya AV, El'Piner IE (1986) Sov. Phys. Accoust. 13: 374
27. a) Winterton RHS (1977) J. Phys. D: Appl. Phys. 10- 204; b) Crum LA (1979) Nature 278: 148
28. Reutskii VV, Starchevskii VL, Makryi EN (1987) Vestn. L'vov. Politekh. Inst. 211: 122
29. Henglein A (1987) Ultrasonics 25: 1
30. Donaldson DJ, Farrington MD, Kruus P (1979) J. Phys. Chem. 83: 3130
31. Weissler A, Pecht I, Anbal M (1965) Science 150: 1288
32. Suslick KS, Gawienowski JJ, Schubert PF, Wang HH (1983) J. Phys. Chem. 87: 2299
33. El'Piner IE (1964) Ultrasound: Physical, chemical and biological effects. F. L. Sinclair trans., Consultants Bureau, New York
34. Regen SL, Singh A (1982) J. Org. Chem. 47: 1587
35. Gautheron B, Tainturier G, Degrand C (1985) J. Am. Chem. Soc. 107: 5579
36. Luche J-L, Petrier C, Dupuy C (1984) Tetrahedron Lett. 25: 753
37. Hansson I, Morch KA, Preece CM (1977) Ultrason. Int. 267
38. de Souza Barbosa JC, Petrier C, Luche J-L (1988) J. Org. Chem. 53: 1212
39. Lauterborn W, Bolle H (1971) J. Fluid Mech. 47: 283
40. Weissler A, Hine EJ (1962) J. Acoust. Soc. Amer. 34: 130
41. Diedrich GK, Kruus P, Rachlis LM (1972) Can. J. Chem. 50: 1743
42. Saskena TK (1980) J. Acoust. Soc. Ind. 8: 12
43. Pugin B (1987) Ultrasonics 25: 49
44. Poddubnyj BN (1976) Sov. Phys. Acoust. 22: 325
45. Lindley J, Mason TJ, Lorimer JP (1987) Ultrasonics 25: 45
46. Rausch MD (1961) J. Org. Chem. 26: 1802
47. Lindström O (1955) J. Acoust. Soc. Am. 27: 654
48. Gueguen H (1963) Ann. Chim. 8: 667
49. Prudhomme RO, Grabar P (1949) J. Chim. Phys. 46: 323
50. Beuthe H (1933) Z. Phys. Chem. 163A(3/4): 161
51. Srivastava SC (1958) Nature 182: 47
52. a) Makino K, Mossoba M, Riesz P (1982) J. Am. Chem. Soc. 104: 3537; b) Makino K, Mossoba M, Riesz P (1983) J. Phys. Chem. 87: 1369
53. Margulis MA, Mal'tsev AN (1968) Zh. Fiz. Khim. 42: 2660
54. Margulis MA (1976) Zh. Fiz. Khim. 50: 1
55. Möckel P (1956) Chem. Tech. 8: 71
56. Kenokh MA (1955) Zh. Obs. Khim. 25: 928
57. Stein G, Weiss J (1948) Nature 161: 650
58. Weissler A (1949) J. Am. Chem. Soc. 71: 419
59. Khenokh M, Lapinskaya EM (1958) Zh. Obs. Khim. 28: 704
60. Siegel G, Pfennigsdorf G, Monig H (1958) Naturwissenschaften 45: 415
61. a) El'Piner IE, Sokol'skaya AV, Margulis MA (1965) Nature (London) 208:

945; b) Starchevskii VL, Vashria TV, Grinder LM, Margulis MA, Mokryi EN (1984) Zh. Fiz. Khim. 58: 1940

62. Pétrier C, Luche J-L (1985) J. Org. Chem. 50: 910
63. Pietrusiewicz KM, Zablocka M (1988) Tetrahedron Lett. 29: 937
64. Renaud P (1950) Bull. Soc. Chim. Fr. Ser. S. 17: 1044
65. Sprich JD, Lewandos GS (1982) Inorg. Chim. Acta. 76: 1241
66. Bönnemann H, Bogdanović B, Brinkman R, He DW, Spliethoff B (1983) Angew. Chem., Int. Ed. Eng. 22: 728
67. Oppolzer W, Schneider P (1984) Tetrahedron Lett. 25: 3305
68. a) Klabunde KJ, Efner HF, Satek L, Donley W (1974) J. Organomet. Chem. 71: 309; b) Oppolzer W, Kündig EP, Bishop PM, Perret C (1982) Tetrahedron Lett. 23: 3901
69. Rieke RD (1977) Acc. Chem. Res. 10: 301; b) Lai Y-H: Synthesis 1981: 585
70. Oppolzer W, Cunningham AF (1986) Tetrahedron Lett. 27: 5467
71. Oppolzer W, Nakao A (1986) Tetrahedron Lett. 27: 5471
72. Bogdanović B, Liao ST, Schwickardi M, Sikorski P, Spliethoff B (1980) Angew. Chem. 92: 845
73. Bogdanović B, Liao S, Mynott R, Schlichte K, Westeppe U (1984) Chem. Ber. 117: 1378
74. Bogdanović B, Schwickardi M, Sikorski P (1982) Angew. Chem. 94: 206
75. Bogdanović B, Schwickardi M (1984) Z. Naturforsch. 39b: 1001
76. a) Brown HC, Racherla U (1986) J. Org. Chem. 51: 427; b) Brown HC, Racherla US (1985) Tetrahedron Lett. 26: 4311
77. Han BH (1985) Daehan Hwahak Hwoejee 29: 557
78. Brettle R, Shibib SM: J. Chem. Soc., Perkin I 1981: 2912
79. de Laszlo SE, Ley SV, Porter RA (1986) J. Chem. Soc., Chem. Commun.: 344
80. Kuchin AV, Nurushev RA, Tolstikov GA (1983) Zh. Obshch. Khim. 53: 2519
81. Yank PH, Lion KF, Lin YT (1986) J. Organomet. Chem. 307: 273
82. Liou KF, Yang PH, Lin YT (1985) J. Organomet. Chem. 294: 145
83. Luche J-L, Pétrier C, Dupuy C (1984) Tetrahedron Lett. 25: 753
84. Einhorn J, Luche J-L (1987) J. Org. Chem. 52: 4124
85. Einhorn C, Allavena C, Luche J-L (1988) J. Chem. Soc., Chem. Commun.: 333
86. Burkow IC, Sydnes LK, Ubeda DCN (1987) Acta. Chem. Scand., Ser B. 41: 235
87. Trost BM, Coppola BP (1982) J. Am. Chem. Soc. 104: 6879
88. Ihara M, Katogi M, Fukumoto K, Kametani T (1987) J. Chem. Soc., Chem. Commun. 721
89. Luche JL, Pétrier C, Gemal AL, Zikra N (1982) J. Org. Chem. 47: 3805
90. Corey EJ, Beames DJ (1972) J. Am. Chem. Soc. 94: 7210
91. Araki S, Butsugan Y (1988) Chem. Lett. 457
92. Naruta Y, Nishigaichi Y, Muruyama K (1986) Chem. Lett. 1857
93. Shirai H, Sato Y, Niwa M (1970) Yakugaku Zasshi 90: 59
94. Han BH, Boudjouk P (1981) Tetrahedron Lett. 22: 2757
95. Boudjouk P, Han BH (1981) Tetrahedron Lett. 22: 3813
96. Scilly NF (1973) Synthesis 161
97. Einhorn J, Luche J-L (1986) Tetrahedron Lett. 27: 1791
98. Einhorn J, Luche J-L (1986) Tetrahedron Lett. 27: 1793
99. Commins ED, Meyers AI (1978) Synthesis 403
100. Einhorn J, Luche J-L (1986) Tetrahedron Lett. 27: 501
101. Boudjouk P, Han BH, Anderson KR (1982) J. Am. Chem. Soc. 104: 4992
102. West R, Fink MJ, Michl J (1981) Science 214: 1343
103. Masamune S, Murakami S, Tobita H (1983) Organomet. 5: 1464

104. Chou TS, Yuan JJ, Tsao CH (1985) J. Chem. Res., Synop. 18
105. Chou TS, Tsao CH, Hung SC (1985) J. Org. Chem. 50: 4329
106. Chou TS, You ML (1985) Tetrahedron Lett. 27: 4495
107. Ley SV, O'Neil IA, Low CMR (1986) Tetrahedron 42: 5363
108. a) Huffman JW, Liao WP, Wallace RH (1987) Tetrahedron Lett. 28: 3315;
     b) Barrett AGM, O'Neil IA (1988) J. Org. Chem. 53: 1815
109 Rautenstrauch V, Willhalm B, Thommen W, Burger U (1981) Helv. Chim.
     Acta. 64: 2109
110. Chou TS, You ML (1987) J. Org. Chem. 52: 2224
111. Chou TS, Chen MM (1987) Heterocycles 26: 2829
112. Bailey WJ, Cummins EW (1954) J. Am. Chem. Soc. 76: 1936
113. Tso HH, Chou TS, Hung SC (1987) J. Chem. Soc., Chem. Commun. 1552
114. Jordan F, Hemmes P, Nishikawa S, Mashima M (1983) J. Am. Chem. Soc.
     105: 2055
115. Azuma T, Yanagida S, Sakurai H, Sasa S, Yoshino K (1982) Synth. Commun.
     12: 137
116. Slough W, Ubbelohde AR (1957) J. Chem. Soc. 918
117. Ley SV, Lygo B, Sternfeld F, Wonnacott A (1986) Tetrahedron 42: 4333
118. Fujita T, Watanabe S, Suga K, Sugahara K, Tsuchimoto K: Chem. and Ind.
     (London) 1983: 167
119. Sugahara K, Fujita T, Watanabe S, Hashimoto H (1987) J. Chem. Tech.
     Biotechnol. 37: 95
120. Kimmel T, Becker D (1984) J. Org. Chem. 49: 2494
121. Xu L, Tao F, Yu T (1985) Tetrahedron Lett. 26: 4231
122. Xu L, Tao F, Yu T (1986) Ziran Zazhi 9: 315; Chem. Abs. 105 (25): 225822
123. Rathke MW (1975) Org. React. 22: 423
124. Rieke RD, Uhm SJ (1975) Synthesis 452
125. White JD, Ruppert JF (1974) J. Org. Chem. 39: 269
126. Rathke MW, Lindert A (1970) J. Org. Chem. 35: 3966
127. Han BH, Boudjouk P (1982) J. Org. Chem. 47: 5030
128. Bose AK, Gupta K, Manhas MS (1984) J. Chem. Soc., Chem. Commun. 86
129. Oguni N, Tomago T, Nagata N (1986) Chem. Express 1: 495
130. Rieke RD (1977) Acc. Chem. Res. 10: 301
131. Pugin B (1986) Ciba-Geigy, Switzerland at the Royal Society of Chemistry,
     Annual Chemical Congress Interdivisional Symposium on Sonochemistry,
     University of Warwick, April 1986. Unpublished results.
132. Brennan J, Hussain FHS (1985) Synthesis 749
133. Aimetti JA, Hamanaka ES, Johnson DA, Kellogg MS: Tetrahedron Lett.
     1979: 4631
134. Ernest I, Gosteli J, Woodward RB (1979) J. Am. Chem. Soc. 101: 6301
135. Carruthers (1982) In: Wilkinson G, Stone FGA, Abel EW (eds) Comprehensive
     organometallic chemistry, Pergamon, vol 7 p 661
136. Bernadou F, Mauze B, Mignac L (1973) C.R. Acad. Sci. Paris Ser. C. 276: 1645
137. Knochel P, Normant JF (1984) Tetrahedron Lett. 25: 1475
138. Repič O, Vogt S (1982) Tetrahedron Lett. 23: 2729
139. Repič O, Lee PG, Giger N (1984) Org. Prep. Proc. Int. 16: 25
140. Yamashita Inoue Y, Kondo T, Hashimoto H (1984) Bull. Chem. Soc. Jpn. 57:
     2335
141. Friedrich EC, Domek JM, Pong RY (1985) J. Org. Chem. 50: 4640
142. Isobe M, Kondo S, Nagasawa M, Goto T (1977) Chem. Lett. 679
143. Luche J-L, Pétrier C, Lansard JP, Greene AE (1983) J. Org. Chem. 48: 3837

144. Pétrier C, Luche J-L, Dupuy C (1984) Tetrahedron Lett. 25: 3463
145. Pétrier C, de Souza Barbosa JC, Dupuy C, Luche JL (1985) J. Org. Chem. 50: 5761
146. Petrier C, Dupuy C, Luche JL (1986) Tetrahedron Lett. 27: 3149
147. Pietrusiewicz KM, Zablocka M, Monkiewicz J (1984) J. Org. Chem. 49: 1522
148. Kitazume T, Ishikawa N: Nippon Kagaku Kaishi 1984: 1725
149. Kitazume T, Ishikawa N: Chem. Lett. 1982: 137
150. Kitazume T, Ishikawa N: Chem. Lett. 1981: 1679
151. Kitazume T, Ishikawa N: Chem. Lett. 1982: 1453
152. Kitazume T, Ishikawa N (1985) J. Am. Chem. Soc. 107: 5186
153. Solladie-Cavallo A, Farkhani D, Fritz S, Lazrak T, Suffert J (1984) Tetrahedron Lett. 25: 4117
154. Han BH, Boudjouk P (1982) J. Org. Chem. 47: 751
155. Chew S, Ferrier RJ: J. Chem. Soc., Commun. 1984: 911
156. Mehta G, Surya Prakash Rao H (1985) Synth., Commun. 15: 991
157. Ghosez L, Montaigne R, Roussel A, Vanlierde H, Mollet P (1971) Tetrahedron 27: 615
158. Hoffmann HMR, Clemens KE, Smithers RH (1972) J. Am. Chem. Soc. 94: 3940
159. Gigure RJ, Rawson DI, Hoffmann HMR: Synthesis 1978: 902
160. Noyori R, Makino S, Takaya H (1971) J. Am. Chem. Soc. 93: 1272
161. Joshi NN, Hoffmann HMR (1986) Tetrahedron Lett. 27: 687
162. Fry AJ, Herr D (1978) Tetrahedron Lett. 40: 1721
163. Fry AJ, Ginsburg GS, Parente RA: J. Chem. Soc., Chem. Commun. 1978: 1040
164. Fry AJ, Bujanauskas JP (1978) J. Org. Chem. 43: 3157
165. Fry AJ, Ginsburg GS (1979) J. Am. Chem. Soc. 101: 3927
166. Fry AJ, Lefor AT (1979) J. Org. Chem. 44: 1270
167. Fry AJ, Donaldson WA, Ginsburg GS (1979) J. Org. Chem. 44: 349
168. Fry AJ, Hong SS (1981) J. Org. Chem. 46: 1962
169. Fry AJ, Ankner K, Hana V (1981) Tetrahedron Lett. 22: 1791
170. Suslick KS, Casadonte DJ, Green MLH, Thompson ME (1987) Ultrasonics 25: 56
171. Chatakondu K, Green MLII, Thompson ME, Suslick KS: J. Chem. Soc., Chem. Commun. 1987: 900
172. Schmidt P, Rosenfeldt E, Milner R, Czerner R, Schellenberger A (1987) Biotechnol. Bioeng. 30: 928
173. Ishimori Y, Karube J, Suzuki S (1981) J. Mol. Catal. 12: 253
174. Bujons J, Guajardo R, Kyler KS (1988) J. Am. Chem. Soc. 110: 604
175. Davidson RS, Patel AM, Safdar A, Thornthwaite D (1983) Tetrahedron Lett. 24: 5907
176. Reddy GS, Smith GG (1987) Inorg. Chim. Acta 133: 1
177. Preston Reeves W, McClusky JV (1983) Tetrahedron Lett. 24: 1585
178. Ezquerra J, Alvarez-Builla J (1985) Org. Prep. Proced. Int. 17: 190
179. Sjoberg K (1966) Tetrahedron Lett. 6383
180. Galin FZ, Ignatyuk UK, Lareev SN, Tostikov GA (1987) Zh. Org. Khim. 23: 1341; (1987) J. Org. Chem. USSR (Engl. Trans.) 23: 1214
181. Mora Ruedas P, Spain ES 549, 102 A1 1st March 1986 (App. 20. 11. 85)
182. Jurczak J, Ostaszewski R (1988) Tetrahedron Lett. 29: 959
183. Mertens J, Vanryckeghem W, Bossuyt A, Van den Winkel P, Vandendriessche R (1984) J. Labelled Compd. Radiopharm. 21: 843

184. For examples see: Schurink HB Org. Synth. Coll. 2: 476
     Ford-Moore AH (1963) Org. Synth. Coll. 4: 84
185. Tashiro M, Nakayama M, Aoki Y, Takigawa A, Maeda K, Tago I, Yoshida M:
     Jpn. Kokai, Tokkyo Koho, JP61/53228 A2586/53228
186. Kajiwara M: Jpn. Kokai, Tokkyo Koho, JP 62/111954 A2[87/111954]
187. Einhorn C, Luche J-L (1986) Carbohydrate Res. 155: 258
188. Schmidt OT (1963) Methods Carbohydr. Chem. 2: 318
189. Christol H, Mousseron M, Plenat F (1959) Bull. Soc. Chim. Franc. 543
190. Krapcho AP, McCullough JE, Nahabedian KV (1965) J. Org. Chem. 30: 139
191. Fujita T, Watanabe S, Sakamoto M, Hashimoto H (1985) Chem. Ind. (London)
     427
192. Nishizawa M, Adachi K, Hayashi Y (1984) J. Chem. Soc., Chem. Commun.
     1637
193. Raucher S, Klein P (1981) J. Org. Chem. 46: 3558
194. Preston Reeves W, McClusky JV (1983) Tetrahedron Lett. 24: 1585
195. a) Galiakhmetov RN, Valitov RB, Kurochkin AK, Margulis MA (1986)
     Zh. Fiz. Khim. 60(4): 1024; b) Galiakhmetov RN, Valitov RB, Kurochkin AK,
     Margulis MA (1985) Zh. Fiz. Khim. 59(12): 2973
196. Palominos MA, Rodriguez R, Vega JC (1986) Chem. Lett. 1251
197. Toma S, Putala M, Salisova M (1987) Collect. Czech. Chem. Commun. 52: 395
198. Ajinomoto Co. Inc., Neth. Appl. NL 85/2968A, 1st July 1986
199. Ninagawa A, Suzuki T, Matsuda H (1986) Chem. Express 1 (3): 169
200. Shibata K, Urano K, Matsui M (1987) Chem. Lett. 519
201. Fuentes A, Sinisterra JV (1986) Tetrahedron Lett. 27: 2967
202. Fuentes A, Marinas JM, Sinisterra JV (1987) Tetrahedron Lett. 28: 2947
203. Fuentes A, Marinas JM, Sinisterra JV (1987) Tetrahedron Lett. 28: 2951
204. a) Artaud I, Seyden-Penne J, Viout P (1980) Tetrahedron Lett. 21: 613; b) Del-
     mas M, LeBigot Y, Gaset A (1980) Tetrahedron Lett. 21: 4831
205. Sinisterra JV, Mouloungi Z, Delmas M, Gaset A (1985) Synthesis 1097
206. Low CMR (1986) PhD Thesis, Imperial College, University of London
207. Priebe H (1984) Acta Chem. Scand., Ser. B, B38: 895
208. Elguero J, Goya P, Lissavetzky J, Valdeomillos AM (1984) C.R. Acad. Sci.
     Paris III 298: 877
209. Davidson RS, Safdar A, Spencer JD, Robinson B (1987) Ultrasonics 25: 35
210. Brown DS, Ley SV, Vile S (1988) Tetrahedron Lett. 29: 4873 (1989)
211. Barot BC, Sullins DW, Eisenbraun EJ (1984) Synth. Comm. 14: 397
212. Varma RS, Kabalka GW (1985) Heterocycles 23: 139
213. Ando T, Kawate T, Yamawaki J, Hanafusa T (1983) Synthesis 637
214. Ando T, Kawate T, Sumi S, Ichihara J, Hanafusa T: Nippon Kagaku Kaishi
     1984: 1731
215. Ando T, Kawate T, Ichihara J, Hanafusa T (1984) Chem. Lett. 72
216. Ando T, Sumi S, Kawate T, Ichihara J, Hanafusa T (1984) J. Chem. Soc., Chem.
     Commun. 439
217. Hanabusa A, Yamaguchi T, Miyamoto K: Jpn. Kokai, Tokkyo Koho,
     JP 62/255466 A2 [87/255466]
218. Casiraghi G, Cornia M, Casnati G, Fava GG, Belicchi MF, Zetta L (1987)
     J. Chem. Soc., Chem. Commun. 794
219. Kawada S, Kurokawa T: Jpn. Kokai Tokkyo JP 62/106073
220. Farooq O, Morteza S, Farina F, Stephenson M, Olah GA (1988) J. Org. Chem.
     53: 2840
221. Schuchardt U, Joekes I, Durate HC (1987) J. Chem. Tech. Biotechnol. 39: 115

222. Menendez JC, Trigo GG, Sollhuber MM (1986) Tetrahedron Lett. 27: 3285
223. Hahn SJ, Kim SH, Chae HJ, Youn BH, Lyu HS (1987) Bull. Korean Chem. Soc. 8: 49
224. Stille JK, Harris FW (1966) J. Het. Chem. 3: 155
225. Thompson DP, Boudjouk P (1988) J. Org. Chem. 53: 2109
226. Naylor EM (1988) PhD Thesis, Imperial College, University of London
227. Doherty AM, Ley SV (1986) Tetrahedron Lett. 27: 105
228. Degrand C (1986) J. Chem. Soc., Chem. Commun. 1113
229. Bard AJ (1963) J. Anal. Chem. 35: 1125
230. Neumann R, Sasson Y (1985) J. Chem. Soc., Chem. Commun. 616
231. Russel GA, Moye AJ, Janzen EG, Mak S, Talaty ER (1967) J. Org. Chem. 32: 137
232. Yamawaki J, Sumi S, Ando T, Hanafusa T: Chem. Lett. 1983: 379
233. Ryoo ES, Shin DH, Han BH (1987) Taehan Hwahakhoe Chi 31: 359
234. Shin DH, Han BH, Cho SY (1986) Daehan Hwahak Hwojee 30: 105
235. Petrier C, Luche J-L (1987) Tetrahedron Lett. 28: 2347 and 2351
236. Sakaii K, Ishige M, Kono H, Motoyama I, Watanabe K, Hata K (1968) Bull. Chem. Soc. Japan 41: 1902
237. Rozantser EG, Shalle VD (1979) Organicheskaya Khimiya Svobodnykh Radikalov, Khimiya, Moscow
238. Kaliska V, Toma S, Lesko J (1987) Collect Czech., Chem. Commun. 52: 2266
239. Lintner W, Hanessian D (1977) Ultrasonics 15: 21
240. Mal'tsev AN (1976) Zh. Fiz. Khim. 50: 1641
241. Shekhobalova VI, Voranova LV (1986) Vestn. Mosk. Univ. Ser 2: Khim. 27: 327
242. Cioffi EA, Prestegard JH (1986) Tetrahedron Lett. 27: 415
243. Popov T, Klisurski D, Ivanov K, Pesheva I (1987) Stud. Surf. Sci. Catal. 31: 191
244. Ivanov K, Popov T, Slavov S (1987) Izv. Khim. 20: 201
245. a) Zembayashi M, Tamao K, Yoshida J, Kumada M (1977) Tetrahedron Lett. 17: 4089; b) Takagi K, Hayama N, Inokawa S (1980) Bull. Chem. Soc., Jpn. 53: 3691; c) Takagi K, Hayama N, Sasaki K (1984) Bull. Chem. Soc.,Jpn. 57: 1887
246. Yamashita J, Inoue Y, Kondo T, Hashimoto H (1986) Chem. Lett. 407
247. Townsend CA, Nguyen LT (1981) J. Am. Chem. Soc. 103: 4582
248. Boudjouk P, Han BH (1983) J. Catal. 79: 489
249. Shin DH, Han BH (1985) Bull. Korean Chem. Soc. 6: 247
250. Han BH, Boudjouk P (1983) Organometallics 2: 769
251. Nakanishi M, JP 81/127,684, Oct. 6. (1981)
252. Mal'tsev AN, Solov'eva IV (1970) Zh. Fiz. Khim. 44: 1092
253. Radenkov F, Khristov K, Kircheva R, Petrov L (1977) Khim. Ind. 49: 11
254. Heck RF, Boss CR (1964) J. Am. Chem. Soc. 86: 2580
255. a) Chen KN, Moriarty RM, DeBoer BG, Churchill MR, Yeh HJC (1975) J. Am. Chem. Soc. 97: 5602; b) Aumann R, Fröhlich K, Ring H (1974) Angew. Chem. Internat. Edn. 13: 275
256. Bates RW, Imperial College, Unpublished Observations
257. Aumann R, Ring H, Kruger C, Goddard R (1979) Chem. Ber. 112: 3644
258. Annis GD, Hebbelthwaite EM, Hodgson ST, Horton AM, Hollinshead DM, Ley SV, Self CR, Sivaramakrishnan R (1984) Second SCI-RSC Medicinal Chemistry Symposium Special Publication No. 50, 148
259. Adams RD, Davison A, Selegue JP (1979) J. Am. Chem. Soc. 101: 7232
260. Cotton FA, Troup JM (1974) J. Am. Chem. Soc. 96: 3438 and 4422

261. Horton AM, Hollinshead DM, Ley SV (1984) Tetrahedron 40: 1737
262. Suslick KS, Schubert PF, Goodale JW (1981) J. Am. Chem. Soc. 103: 7342
263. Begley MJ, Puntambekar SG, Wright AH (1987) J. Chem. Soc., Chem. Commun. 1251
264. Manuel TA (1964) Inorg. Chem. 3: 1794
265. Annis GD, Ley SV, Self CR, Sivaramarkrishnan R, Williams DJ (1982) J. Chem. Soc., Perkin Trans I 1355
266. Suslick KS, Goodale JW, Schubert PF, Wang HH (1983) J. Am. Chem. Soc. 105: 5781
267. a) Annis GD, Ley SV (1977) J. Chem. Soc. Chem. Commun. 581; b) Annis GD, Ley SV, Self CR, Sivaramakrishnan R (1981) J. Chem. Soc., Perkin Trans I 270
268. Horton AM, Ley SV (1985) J. Organomet. Chem. 285: C17
269. Organ HM (1987) PhD Thesis, Imperial College, University of London
270. White AD (1988) PhD Thesis, Imperial College, University of London
271. Annis GD, Hebblethwaite EM, Hodgson ST, Hollinshead DM, Ley SV (1983) J. Chem. Soc., Perkin Trans. I 2851
272. a) Haug T von, Lohse F, Metzger K, Batzer H (1968) Helv. Chim. Acta 51: 2069; b) Moriconi EJ, Mayer WC (1971) J. Org. Chem. 36: 2841
273. Hodgson ST, Hollinshead DM, Ley SV (1985) Tetrahedron 41: 5871
274. Hodgson ST, Hollinshead DM, Ley SV, Low CMR, Williams DJ (1985) J. Chem. Soc. Perkin Trans. I 2375
275. Pettit R, Emerson GF (1964) Adv. Organomet. Chem. 1: 1
276. Pearson AJ (1982) In: Wilkinson G, Stone FGA, Abel EW (eds) Comprehensive organometallic chemistry, Pergamon, Oxford, vol 8 p 939
277. Ley SV, Low CMR, White AD (1986) J. Organomet. Chem. 302: C13
278. Cais M, Maoz N (1966) J. Organomet. Chem. 5: 370
279. Emerson GF, Ehrlich K, Giering WP, Lautebur PC (1966) J. Am. Chem. Soc. 88: 3172
280. Suslick KS, Johnson RE (1984) J. Am. Chem. Soc. 106: 6856
281. a) Werner RPM, Filbey AH, Manastryskj SA (1964) Inorg. Chem. 3: 298; b) Ellis JE, Davison A (1976) Inorg. Synth. 16: 68
282. Suslick KS, Schubert PF (1983) J. Am. Chem. Soc. 105: 6042

Subject Index